D1318176

Biological
Identification

The principles and practice of identification methods
in biology

Richard J. Pankhurst

Department of Botany
British Museum (Natural History)

University Park Press
Baltimore

First published 1978
by Edward Arnold (Publishers) Limited, London

Published in the U.S.A. by
University Park Press
233 East Redwood Street
Baltimore, Maryland 21202

Library of Congress Cataloguing in Publication Data

Pankhurst, R. J
  Biological identification.

  Includes bibliographical references and index.
  1. Biology—Classification. I. Title.
QH83.P36 1978 574'.01'2 78–11187
ISBN 0–8391–1344–7

Printed in Great Britain

'What's the use of their having names,' the Gnat said, 'if they won't answer to them?'

'No use to *them*,' said Alice; 'but it's useful to the people that name them, I suppose. If not, why do things have names at all?'

Through the Looking Glass
Lewis Carroll

For Anne, Lucie and Thomas.

Preface

This book is intended for all university students of biology, and for professional biologists who need to have some working knowledge of identification methods. It is probably the first book of its kind on this subject, as the identification of specimens is usually treated as a small section of larger textbooks on taxonomy, in spite of its considerable practical importance. There can hardly be any biologist who follows his or her career without at some stage needing to know the identity of some specimens of animals or plants which are being studied. There must be many whose daily business involves identifying specimens, but whose knowledge of taxonomy is only a means, and not an end in itself. Hence this book puts much emphasis on the practical aspects of identification methods.

An account is given of both the traditional and modern methods, and in the latter the use of a computer is often necessary. The one method with which most readers are likely to be familiar is the diagnostic key, and this is discussed in detail. On the other hand, one's view of the subject would be distorted without a proper understanding of all the other methods which have been developed in recent years. Many of these advances were only possible with the use of computers. However, those who are nervous of becoming involved with the computer ought to realize that many of the methods only require computation at a preparatory stage, and not for daily use. A short history of the subject is presented, and a brief chapter on some selected applications is provided. There is relatively little mathematics, and what there is is quite straightforward. Statistical methods are treated briefly, since practical difficulties outweigh their theoretical advantages. This is not the place to go into the details of the use of computer programs, or into the principles of computer science, and such matters are only explained in general terms. While I, personally, confess to being a botanist, I have gone to a good deal of trouble to give a balanced account, suitable for biologists of all persuasions.

This book complements the following titles which have already been published by Edward Arnold, namely 'Plant Taxonomy', by V. H. Heywood, Institute of Biology, Studies in Biology, No. 5, and 'Biological Nomenclature', by Charles Jeffrey.

I wish to thank my family for their patience during the days when I shut myself away at home in order to write. I have a long-standing debt to Max Walters, Director of the Botanic Garden at Cambridge, for encouraging me to start research on identification methods in the first place, back in 1968, and for helping me ever since. My colleagues in the various departments of the British Museum (Natural History) have helped by numerous discussions over the past two years, and I particularly want to thank John Cannon, Keeper of Botany, for prompting me to start writing this book and for criticizing the text, and Paul Whalley, Department of Entomology, for reading and criticizing it too. Paul Cannon, University of Reading, and Bob Allkin, Polytechnic of Central London, also made useful comments on the manuscript. Individual acknowledgements for permission to reproduce certain of the figures are as follows:-

Fig. 6 Michael Chinery, and Collins Ltd.
Fig. 9 Paul Whalley, and the Copeland-Chatterson Co.
Fig. 10 Drs. Hansen and Rahn, Botanical Museum, Copenhagen.
Fig. 11 Charles Sinker, Director of the Field Studies Council.
Fig. 12 Dr. Nash, Western Hospital, London.
Fig. 14 Paul Whalley.
Fig. 15 Prof. Gyllenberg, of the Academy of Finland, the Systematics Association, and Academic Press.
Fig. 16 S. P. Lapage and W. R. Willcox, Central Public Health Laboratory, London.
Figs. 19 and 20 British Museum (Natural History).

R. J. Pankhurst

British Museum (Natural History)
1978

Contents

1 General Introduction

1.1 Background

The identification of objects is a fundamental human activity. We practice it, for example, when we read the printed word, or recognize someone we know. For a biologist, identification usually means finding the name for a specimen of animal or plant, and the specimen to be identified is usually assigned to a species. Whatever sort of object is in question, it cannot be identified unless there is already a classification of like objects with which the new object can be compared. Classification, here, means a way of grouping objects, on the basis of some relationship between them. The groups so formed are very often given names, and when a new object is examined, and it is decided that it belongs to one of the existing groups, then it has been identified. The purpose of these remarks is to make clear that the word 'classification' is being used in a special sense. In ordinary English, 'classify' can mean both its special sense as just given, and also it can mean 'to identify'. The verb 'to recognize' has these two meanings also. Biologists often also speak of the 'determination' of specimens, which means the same as identification, and of the 'naming' of specimens, which means identification too, but sometimes implies that this has been done temporarily or superficially.

By far the greatest part of the information which a biologist uses when identifying specimens is based on gross morphology, that is, the features that can be observed with ease, such as shape, colour and size, when the object is held in the hand. This is not because such facts are necessarily fundamental in any sense, but simply for practical reasons of speed and convenience. There is a strong tendency to avoid more specialized techniques such as dissection or microscopy, even though they may yield valuable information. Although automatic data-collecting techniques exist, most of the information used in identification is obtained by human observation and interpretation, and this seems likely to remain the case for the time being.

The diagnostic key, or 'key' for short, is by far the most frequently used identification method, and many biologists will not need an explanation of what this is. For those not familiar with it, an explanation of its use and construction is given on p. 11. The key is used in most fields of biology, and its use is several centuries old. Various modern techniques are now known, which are not yet commonly applied, but which have much to commend them. This book will review both traditional and modern methods. The word 'key' is also loosely applied to identification methods other than the diagnostic key.

1.2 Some definitions

The fundamental item of information is called a *character*. For example, 'flower colour' is a character for a plant. Many other words are used for the same thing e.g. characteristic, feature, property, attribute, symptom, sign, facet and test result.

Characters when observed are seen to show various *states* (also called *values* or *attributes*). For example, the states of 'flower colour' might be 'red', 'white', and 'blue'. The term 'attribute' has been used variously for both 'character' and 'state', so its use is not recommended. A character may be *constant*, if the object(s) in question have the same one and only state in all cases. If several states of a character are observed on one kind of organism, the character is *variable*. Notice that what is considered constant depends on what objects are being discussed, so this is a relative concept.

Where biology is concerned, the objects to be identified will mostly be species, or genera, or other such groupings at various levels. These are in general called *taxa* (singular *taxon*).

Characters and states are frequently combined in phrases such as 'petals white', and such an expression, consisting of a character ('petal colour') with one of its states ('white') is often loosely referred to as a 'character'. This distinction may seem rather a fine one, but tends to cause confusion when preparing descriptions of taxa for processing by computer. It may help to remember that a character is a noun or noun phrase, and that a state is an adjective or adjectival phrase. In what follows, 'character' will always be used in the strict sense.

Fundamental to any identification scheme is a summary of the classification on which it is based. This may be a series of written descriptions of the taxa, but is conveniently expressed as a table of taxa and characters, with the states filled in (Fig. 1). This is often referred to as a *taxonomic data matrix*, or just a *data matrix*. The mathematical term 'matrix' is only borrowed, and has little theoretical significance. The matrix is in general a rectangular one i.e. the number of rows is not the same as the number of columns. Whether the matrix is drawn up with taxa in rows or in columns is immaterial.

For a more complete glossary of technical terms used in identification, see Morse *et al.* (1975).

1.3 Types of identification method

There are two principal kinds of method, in a broad sense, and this is true of any identification problem.

(i) Monothetic. This means that only one character is used at a time. The familiar diagnostic key is an example of this. Characters are used, one at a time in sequence. The identification is often achieved with only a proportion of all the characters which are available on the specimen. The specimen should fit exactly to the description of the taxon with which it is identified. Such identification methods may be *single-access* (with only one sequence and choice of characters which can be used for each specimen) or *multi-access* (with any choice or sequence of characters).

(ii) Polythetic. This means that the method uses several characters simultaneously. An example is the tabular key (p. 33). Usually the specimen is described fairly completely before attempting identification, but it is not essential that every character of the specimen should agree with the taxon with which it is being identified.

This distinction between monothetic and polythetic methods is not hard and fast, as there are intermediate techniques (p. 73).

1.4 Hierarchies, trees and keys

Most classifications which biologists find useful are hierarchical. They consist of groups of taxa at different levels, some more general than others. Each level corresponds to a taxon e.g. a family or genus. Each taxon can belong to only one taxon on the next higher level. This is essential if taxa are going to be identified, since an identification gives a name for labelling and reference purposes, and this is not very useful if it is ambiguous. A hierarchy can be drawn in the form of what mathematicians call a *tree.* This tree is usually drawn with its root at the top and its branches at the bottom! Figure 2 illustrates a hierarchy (a classification) drawn as a tree. In this tree boxes represent taxa, and the lines show the hierarchical relationships between them e.g. each species is connected by a line to the genus to which it belongs. By contrast, an example of a non-hierarchical classification can be given, namely that of fruit and vegetable, as in everyday use. The banana is either a fruit or a vegetable, depending on how it is used, so here is a 'taxon' which belongs to two higher groups at once, namely 'fruit' and 'vegetable'.

It is important to realize that any hierarchy can be drawn as a tree. One example already given is a classification, where we have deliberately not said anything about what sort of a classification is being represented, since there are various possibilities. One popular form of

Fig. 1. A data matrix for British species of *Epilobium*.

CHARACTERS \ SPECIES	1 *hirsutum*	2 *parviflorum*	3 *montanum*	4 *lanceolatum*	5 *roseum*	6 *adenocaulon*	7 *tetragonum*	8 *lamyi*	9 *obscurum*	10 *palustre*	11 *anagallidifolium*	12 *alsinifolium*	13 *brunnescens*
1 Habit of plant	E	E	E	E	E	E	E	E	E	E	D	D	P
2 Stem simple hairs	H	G or H	G or H	H	H	H	H	H	H	G or H	G	G	
3 Habit of stem simple hairs (if present)	S	S	A	A	A	A	A	A	A	A	*	*	
4 Stem glandular hairs	+	+	−	−	+	+	−	−	−	−	−	−	−
5 Stem lines	T	T	T	L	L	L	L	L	L	T	L	L	L
6 Stem rooting at nodes	−	−	−	−	−	−	−	−	−	−	−	−	+'
7 Stem leaves all sessile (without stalks)	+	+	−	−	−	−	+	−	+	±	+	−	±
8 Leaves amplexicaul (clasping stem)	semi	−	−	−	−	−	−	−	−	−	−	−	−
9 Leaves decurrent (running down stem)	+	−	−	−	−	−	+	+	+				

Character	1	2	3	4	5	6	7	8	9	10	11	12	13
10 Leaf base shape	C	R	C	C	R	C	C	C	R	C	C	R	R
11 Leaf shiny	+	+	+	−	+	−	+				−	+	
12 Flower position	Te	Te	Te	Te	Te	Te	Te	Te	Te	Te	Te	Te	Ax
13 Flower diameter (mm)	>10	6–10	6–10	<6	6–10	>10	6–10	6–10	<6	<6	6–10	6–10	<6
14 Flower colour	Ro	Pi	Pi	Pi	Wp	Pi	Pi	Pi	Ro	Pi	Ro	Pi	Pi
15 Stigma lobed	+	+	+	−	−	−	−	−	−	−	−	−	−
16 Stigma length relative to style			∨	∨	⩽	⩽	∨	⩽	⩽	∨			
17 Glandular hairs on 'calyx tube'			+	−	+	−	+		+				
18 Fruit stalk length (cm)	<2	<2	<2	<2	<2	<2	<2	<2	<2	<2	<2	2–5	2–5

Abbreviations:

A	appressed, lying against stem
Ax	axillary (in leaf axils, where leaves and stem join)
C	cuneate (wedge-shaped)
D	decumbent to ascending i.e. stem follows ground at first, or rises at an angle
E	more or less erect i.e. upright
G	(sub) glabrous (more or less without hairs)
H	hairy
L	with raised lines
P	prostrate
Pi	pink
R	rounded
Ro	rose
S	spreading, standing out from stem
T	more or less terete i.e. round, without lines
Te	terminal
WP	white to pale pink
∨	shorter, less than
⩽	about equal
∧	longer, greater than
*	inapplicable
+	yes
−	no

Fig. 2. Classification of some native British *Onagraceae*, including genus *Epilobium*, drawn as a tree.

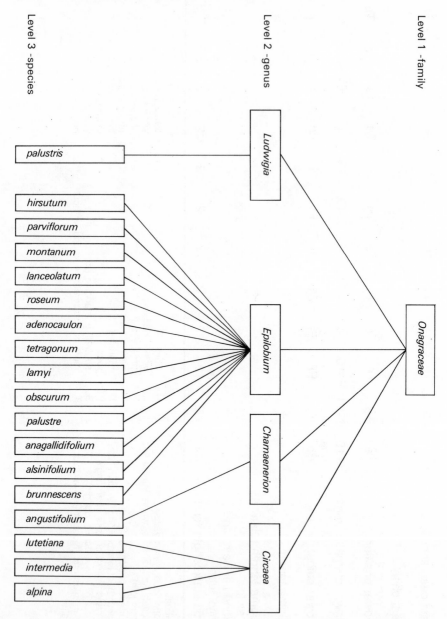

classification is that based on ancestry by evolution, known as a *phylogenetic* classification and in this case the tree representing the ancestry (called a *cladogram*) is equivalent to the classification. On the other hand, many classifications are based on observations of characters of specimens alone, and hence on their overall present-day resemblance. Such classifications are said to be *phenetic*. A tree diagram of a phenetic classification therefore does not necessarily correspond to an ancestry, although it may do so. It is sometimes called a *dendrogram*. There has been much controversy over whether the phylogenetic or phenetic approach is the proper one to use, but this will not be discussed here. For further reading, contrast the views of Sneath and Sokal (1974) with those of Hennig (1966).

Lastly, the popular identification key, or diagnostic key, can also be drawn as a tree, where the boxes represent the decisions to be taken, and only the boxes on the final branches represent taxa. Figure 3 represents the same key as in Figs. 4 and 5, but expressed as a tree. Hence the tree of a key is something different again. Some authors like to construct keys which reflect the classification, and these are called *natural* keys. The tree of a key made this way may look very much like the dendrogram or cladogram on which it is based, as characters or combinations of characters can always be chosen from the taxa in order to do this. However, such keys tend to require the use of several characters in combination, any or all of which are often qualified by adjectives such as 'usually' and 'sometimes'. As a result, a natural key is harder to use and less reliable than the *artificial* key. An artificial key is based on whatever characters are easiest to observe and which produce the answer most quickly and reliably. It therefore need not bear any obvious resemblance to the classification to which it corresponds.

The natural key is a way of setting out a classification and it is not recommended simply for the practical purpose of obtaining fast, reliable identification. From this point onward, by 'key' we shall understand 'artificial key'. It is evident that when one wishes to identify a specimen, one must proceed with the characters it presents on the basis that no knowledge is available at that moment of the phylogenetic relationships of the species to which it belongs. In this sense, all identifications are necessarily phenetic. It is possible for a species to evolve in such a way that it has a close but superficial resemblance to another species to which it is not very closely related by ancestry. This is known as *convergent evolution*. A phylogenetic classification involving two such species should separate them to a degree which reflects their origin, and not their present-day similarity. On the other hand, an efficient key for identifying these two species must deal with them together, since they are rather alike. It is also

Fig. 3. Simplified *Epilobium* key drawn as a tree.

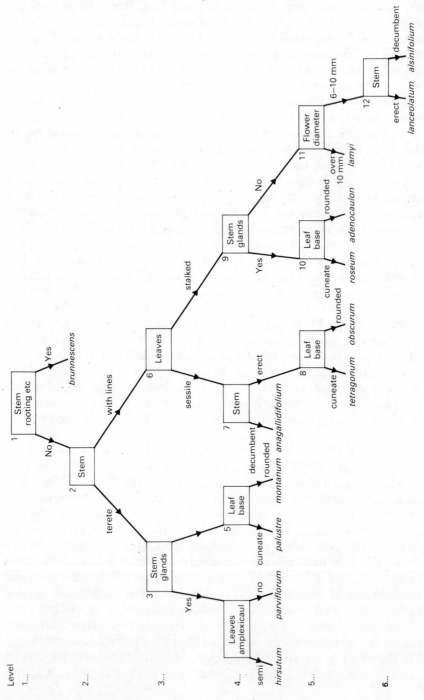

possible that the characters which are thought to be significant for phylogeny and which are presented in a classification are not the same as those which are convenient for identification. These remarks show how principles of classification can conflict with the practical needs of identification. For the purposes of the remainder of this book, the classification of a group of organisms will be taken for granted, without regard to methods by which it is created, as a starting point for identification problems.

1.5 Pattern recognition

Pattern recognition is a general term for the application of computers to problems of recognition. The relation of this subject to the identification of biological specimens is briefly discussed in this section.

Pattern recognition methods are applied to a great diversity of problems, including biological applications. As a broad generalization, one can say that much pattern recognition work has an extra component which is missing from the conventional practice of identification by biologists, because the object to be recognized is very frequently *digitized* at a preliminary stage. This means that electronic equipment (such as a television camera, or a pencil follower) is first used to obtain a representation of the object of interest. This representation is sometimes a table of values of lightness or darkness over an area, or a sequence of geometrical coordinates for positions on a line. In other words, the description of the object to be identified is obtained automatically in a form that a computer can handle by automatic processes. This is very much in contrast with a biologist identifying a specimen, who will mostly make his or her own observations, and then draw conclusions, without much direct use of any equipment (except perhaps a microscope).

Once the description of an object has been stored in a computer, the stage of *feature extraction* still has to take place. This is the equivalent of recognizing the shape of a leaf of a plant, or counting the number of antennal segments on an insect. What a human observer perceives in a few seconds can be a complex process for a computer to perform. A further difficulty is the physical complexity of natural organisms, with the variations of growth patterns in any one species, and in the problems of dealing with an object in three dimensions seen from an arbitrary direction. The more successful uses of computers in pattern recognition often concern objects which are relatively simple, e.g. only objects with regular edges and surfaces in the analysis of pictures, or which have only two dimensions, such as the karyology of human chromosomes, or recognition of hand-writing. For a general introduction to pattern recognition see Mendel and Fu (1970). For the time being, it seems likely that most identifications of plants and animals will depend on human observation (but see Chapter 3, p. 78).

This is not the place to go into the theory of pattern recognition, but some useful comments can be made. For any polythetic identification method which calculates a measure of similarity by adding to the measure for each character which agrees, it is possible to use a numerical weighting for each character in the hope of improving the results (see p. 56). There is a theorem which shows that if such a method is used, and if there exists a set of weights which will give the correct identification with a set of samples, then it is always possible to compute those weights. This result is known as the *perceptron convergence theorem* (Minsky and Papert, 1969). Put in other words, if correct identification by such a method is possible, then the necessary character weights can always be calculated. One can find a set of examples and calculate the weights, and then use the weights to identify more samples, calculate the weights again, and so on. The significance of this is that it is a learning process, and if carried out by computer, what is taking place is a process of *artificial intelligence*, since a computer program has altered its behaviour as a result of experience. This is one of the important aspects of human intelligence.

2 Conventional Identification Methods

In this chapter, all identification methods which make no use of calculation or computers are discussed. These are, broadly speaking, the methods most used up to the present time: the diagnostic key, the multi-access key, and the comparison method.

The discussion in this chapter, and elsewhere, is unified to some extent by the repeated use of the same example, the British species of the higher plant genus *Epilobium*. The descriptive data in Fig. 1 are largely derived from Clapham, Tutin and Warburg (1962). For non-botanists, such technical terms as are used are defined below Fig. 1. Figs. 2, 3, 4, 5 and 7 were all derived from Fig. 1 by the author.

2.1 Diagnostic keys

2.1.1 Use

There are two principal kinds of key: the *parallel* (or *bracketed*) style, as in Fig. 4, and the *yoked* (or *indented*) style, as in Fig. 5. The term 'indented' is unfortunate, because indentation could be used in the layout of either kind. The essential difference is that in the parallel variety, all the components of a set of related questions are printed together, whereas in the yoked kind, all the possibilities which depend on part of a question are set down and exhausted first before other parts of a question are taken up. For example in Fig. 4 (parallel), the two kinds of 'leaf stalk length' i.e. 'sessile' or 'stalked' appear together as the components of the question labelled 6. On the other hand, in Fig. 5 (yoked), the first part of question 6 (leaves sessile) is fully explored, down to the question labelled 8, before the alternative (leaves stalked) is taken up again. These alternative parts of questions in keys are called *leads* or *legs* (because they lead from one question to another).

Before discussing how to use a key, one needs to consider the matter of finding or choosing a key. Often the identity of a specimen in the wider sense is already known e.g. there may be no difficulty in knowing that one has an insect, or a fern. If the class, order or family is uncertain, there exist keys for these broad categories e.g. for families

Fig. 4. Key to British *Epilobium* species, parallel style.

1	Stem rooting at nodes, prostrate, flowers axillary.	*brunnescens*
	Stem not rooting at nodes, decumbent, ascending or erect, flowers terminal	2
2(1)	Stem more or less terete	3
	Stem with raised lines	6
3(2)	Stem with glandular hairs	4
	Stem without glandular hairs	5
4(3)	Leaves semi-amplexicaul and decurrent, cuneate at base, flowers over 10 mm diam., petals rose	*hirsutum*
	Leaves not amplexicaul, nor decurrent, rounded at base, flowers 6–10 mm diam., petals pink	*parviflorum*
5(3)	Leaf cuneate at base, flowers under 6 mm diam., stigma entire	*palustre*
	Leaf rounded at base, flowers 6–10 mm diam., stigma 4-lobed	*montanum*
6(2)	Leaves more or less sessile	7
	At least some leaves stalked	9
7(6)	Stem decumbent to ascending, subglabrous, flowers under 6 mm diam., fruit stalk 2–5 cm	*anagallidifolium*
	Stem more or less erect, with simple hairs, flowers 6–10 mm diam., fruit stalk under 2 cm	8
8(7)	Leaf cuneate at base, shiny above, glandular hairs absent from 'calyx tube', petals pink	*tetragonum*
	Leaf rounded at base, dull above, glandular hairs present on 'calyx tube', petals rose	*obscurum*
9(6)	Stem with glandular hairs, flowers under 6 mm diam.	10
	Stem without glandular hairs, flowers over 6 mm diam.	11
10(9)	Leaf cuneate at base, petals white to pale pink, stigma more or less equal to style	*roseum*
	Leaf rounded at base, petals pink, stigma less than style	*adenocaulon*
11(9)	Flowers over 10 mm diam.	*lamyi*
	Flowers 6–10 mm diam.	12
12(11)	Stem more or less erect, with simple hairs, leaf cuneate at base, stigma 4-lobed, fruit stalk under 2 cm	*lanceolatum*
	Stem decumbent to ascending, subglabrous, leaf rounded at base, stigma entire, fruit stalk 2–5 cm	*alsinifolium*

of angiosperms (flowering plants), (Davis and Cullen 1965), or for orders of insects in the world (Brues *et al*. 1954). Textbooks on taxonomy contain general advice on identification, e.g. for plants (Lawrence 1951), for animals (Mayr 1969) and for microbes (Cowan and Steel 1965). It may be possible to find a key which is restricted to the appropriate geographical area, so that there is no need to consider taxa which are not known to occur in the area of origin of the specimen. Even if there is no such key, there may exist a check list of known taxa from the area in question, and this can be used in conjunction with a more generalized key to help eliminate some of the possibilities.

Let us illustrate the use of a key with Fig. 4 (a parallel key). We have a specimen of a plant to identify. There must be some reason for choosing to use this key i.e. we are assuming that the plant is (a) from the genus *Epilobium* (possibly by using a generic key before), and (b) that it is one of those included in this key (e.g. because it came from the appropriate geographic area). If we are confident that these conditions are met, then begin at the top with the question labelled 1 at the left, i.e. does the stem root at the nodes, etc? Suppose the answer is no, and the stem is erect, then the number of the next question is shown at the far right, i.e. 2. Refer now to question 2, i.e. is the stem terete (round) or does it have raised lines? If raised lines are present, we lead on to question 6. Continuing this process, assuming that the specimen has stalked leaves, and glandular hairs on the stem, and leaf rounded at base, we arrive at the second lead of question 10. At the right there is the name of a species (*adenocaulon*) instead of the number of a further question. Our specimen has 'keyed out' to *E. adenocaulon*. To be sure this is right, check the characters given in the rest of the lead (if any), and then look up a complete description of the taxon and compare, or else compare it with named museum specimens, line drawings or photographs, as convenient. The identification should usually also agree with one's previous experience, and fit in with knowledge of geographical distribution and of seasonality.

The use of the yoked key of Fig. 5 is little different from this. Since the two leads of question 2 are not printed together, we have to look down the page for the alternative leads. This is why the questions are indented, so as to make the related leads easier to find. The last lead of a question is marked with an asterisk, so that none is overlooked in cases where there are more than two. Once we have chosen a lead, e.g. 'stem with raised lines', the next question is immediately in the next line i.e. number 6, 'leaves more or less sessile'.

What has been said so far about using a key has assumed that there were no errors in the recognition of characters of the specimen. Suppose however, using the above example again, that we really did

Fig. 5. Key to British *Epilobium* species, yoked style.

1	Stem rooting at nodes, prostrate, flowers axillary	*brunnescens*
1*	Stem not rooting at nodes, decumbent, ascending, or erect, flowers terminal	2
2	Stem more or less terete	3
3	Stem with glandular hairs	4
4	Leaves semi-amplexicaul and decurrent, cuneate at base, flowers over 10 mm diam., petals rose	*hirsutum*
4*	Leaves not amplexicaul, not decurrent, rounded at base, flowers 6–10 mm diam., petals pink	*parviflorum*
3*	Stem without glandular hairs	5
5	Leaf cuneate at base, flowers under 6 mm diam., stigma entire	*palustre*
5*	Leaf rounded at base, flowers 6–10 mm diam., stigma 4-lobed	*montanum*
2*	Stem with raised lines	6
6	Leaves more or less sessile	7
7	Stem decumbent to ascending, subglabrous, flowers under 6 mm diam., fruit stalk 2–5 cm	*anagallidifolium*
7*	Stem more or less erect with simple hairs, flowers 6–10 mm diam., fruit stalk under 2 cm	8
8	Leaf cuneate at base, shiny above, glandular hairs absent from 'calyx tube', petals pink	*tetragonum*
8*	Leaf rounded at base, dull above, glandular hairs present on 'calyx tube', petals rose	*obscurum*
6*	At least some leaves stalked	9
9	Stem with glandular hairs, flowers under 6 mm diam.	10
10	Leaf cuneate at base, petals white to pale pink, stigma more or less equal to style	*roseum*
10*	Leaf rounded at base, petals pink, stigma less than style	*adenocaulon*
9*	Stem without glandular hairs, flowers over 6 mm diam.	11
11	Flowers over 10 mm diam.	*lamyi*
11*	Flowers 6–10 mm diam.	12
12	Stem more or less erect, with simple hairs, leaf cuneate at base, stigma 4-lobed, fruit stalk under 2 cm	*lanceolatum*
12*	Stem decumbent to ascending, subglabrous, leaf rounded at base, stigma entire, fruit stalk 2–5 cm	*alsinifolium*

have a specimen of *E. adenocaulon*, but erroneously thought that its leaves were sessile, leading to question 7. Here the second lead fits much better than the first, although the flower diameter is not correct. This then leads to 8, and 'leaf rounded at base' suggests *obscurum*. A check with a Flora should show that this is wrong, because the stem of *obscurum* has no glandular hairs, whereas *adenocaulon* has, besides differences of flower size and colour (see Fig. 1). Failure to check might result in the mistake passing unnoticed! Once the mistake is recognized one can return to the key and work back from the lead for *E. obscurum*, and ask which character might have been wrong, and when a mistake is found, work forwards through the key again to another answer. It is often said that the yoked type of key is better for working backwards in this situation, because the previous lead, which pointed to the current one, is always on the line above. However a parallel key can easily be provided with the lead numbers of previous questions. In Fig. 4, these are provided in brackets after the question numbers on the left.

Another situation where errors can arise is when the specimen is of a taxon not included in the key. It might be a new species, a non-native or a hybrid, or, most likely, belong to another genus or family altogether. Once again, it is possible to work right through the key without noticing anything wrong, unless the tentative identification is checked afterwards. Incongruities in characters can occur which will show up this situation. For example, in question 2 of the *Epilobium* key, 'stem terete' (rounded in cross-section) is compared with 'stem with raised lines'. If the specimen had a square stem, an error would at once be suspected.

Suppose that one were not sure whether the leaves had stalks or not. One could temporize, by trying all outcomes of question 6, i.e. questions 7 and 9; but if question 7 again raises a doubt, confusion is soon reached. This illustrates a fundamental fault of the key as an identification method; namely, that it requires the use of specified characters which may not be available or convenient. To some extent, this difficulty can be avoided by providing alternative keys, e.g. for flowering and fruiting material, or for male as well as female insects, but it is impossible to construct a diagnostic key which will meet every situation.

2.1.2 Construction

Any key has to be based on a knowledge of characters and their corresponding states. This is obtained by examining living organisms, preserved specimens and taxonomic handbooks. There are hazards even in this process, since specimens (living or preserved) may not be correctly identified, and different authors may not use the same definitions of taxa, or may be using the same descriptive terms with different meanings.

The different characters, once obtained, will differ in their usefulness for making keys. In what follows, a 'good' character means one which is good for making keys, and is not necessarily useful for other purposes. An assessment of the 'goodness' of characters is often called *character weighting*. This may be expressed as a preference, or numerically. Good characters have various aspects:

(i) Ease of observation. Characters which can be seen with the unaided eye will always be more practical to observe than those requiring special equipment, e.g. microscopy or chemical tests. Under this heading also come questions of availability of characters, e.g. for a species with two sexes a character which occurs on both is better. Similarly, for preserved specimens, colours are often lost and some characters are seasonal in their appearance, e.g. those of the fruit of plants.

(ii) Information content. Suppose there are 10 species in a genus, of which five show a certain state of a character and five show another. This is then a very informative character for separating species. On the other hand, states in the proportion of 1 to 9 denote a character which gives little information. Such a character may be highly *diagnostic* for a particular species, but is little use for distinguishing species in general. Another way to put this is that a character which is very variable *between* species is highly informative. On the other hand, a character which varies greatly *within* species and whose states overlap, is very little use for distinguishing taxa and carries little information. Finally, a character which is constant throughout the genus has no information content at all, except for defining the genus, but might be useful for distinguishing it from other genera. The information content of a character has meaning only in relation to the taxa which are being considered, and is not an absolute concept.

The definition of characters and states which are being used may be perfectly evident from standard usage, but if any terms are being used in a non-standard way, or if specialized characters occur which apply only to the organisms in question, then it is important to supply definitions. For the example which follows, definitions are supplied under Fig.1. Some well-chosen line drawings or illustrations may be very helpful here.

Characters can be of different types. A *qualitative* character is simply a statement, such as 'shape of head' whereas a *quantitative* character concerns a number or a measurement, e.g. 'number of legs', 'height of stem'. Quantitative characters may be *discrete*, e.g. 'number of petals' or *continuous*, e.g. 'length of wing'. Continuous quantitative characters tend to show a great deal of variation, mostly due to environmental influences on the growth of organisms, and are therefore often poor characters from the point of view of information content.

It might seem unnecessary to point out that states of characters used in keys have to be *contrasting*, or *mutually exclusive*, were it not for the number of examples available where this rule is broken. For example, suppose the character 'petal colour' has been given three states, 'red', 'blue' and 'spotted'. There are really two characters here, namely: 'colour of petal' (= red or blue) and 'distribution of petal colour' (= uniform, or spotted). A way to detect this failure to contrast is to ask whether any two states can apply simultaneously, e.g. can petals be 'red' and 'spotted' at the same time? They can, so these states are not exclusive and do not contrast. If states are not exclusive of each other, and are used in a key, it may be impossible to decide which lead to take. Not only must states be mutually exclusive, so must leads. For example:

> Wings present, 5–12 mm long.
> Wings absent or present, and if present, 8–20 mm long.

will be useless for a winged specimen with wings 10 mm long.

The starting point for the construction of any key is a rectangular table of the taxa and their characters. This is also called a *data matrix*, and an example is shown in Fig. 1. There are a number of gaps in this table where characters are missing. The reason for this is simply that the table is based on a source which did not contain all the information. The source would have been more useful for this purpose if the information had been complete, because greater flexibility in key construction would then have been possible. Another reason why characters may not be scored is that it may be impossible to score them. For example, if insects in a group can be winged or wingless, then 'shape of wing' is a character which is irrelevant for wingless species. This is called a *conditional* (or *dependent*) character.

Which type of key is to be preferred, parallel or yoked? The former may take slightly less space in publications, because no room is taken up for indentation. If suitably presented, they are both equally suited to back-tracking for errors. The yoked type is said to be less convenient when it comes to finding contrasting leads, but this can be mitigated by providing marginal notes to show where the next lead is (if it is on another page), and whether it is the final lead of its kind. On the other hand, the yoked type can show the relationships between species, in so far as a key is a suitable way of doing this. It has already been argued (p. 7) that this consideration is not relevant. Hence there is very little difference in merit on technical grounds between the two kinds of key, although there may be subjective preferences. This viewpoint is rather in contrast with that of other authors, who frequently argue strongly for one kind or the other.

We now come to the construction of an actual key. To begin with, choose a character which is easy to observe and understand, and which divides the species into two groups of as nearly as possible equal size. The advantage of equal branching is that it tends to produce a key with the smallest average number of questions required to reach each taxon; and with fewer questions, other things being equal, there are fewer chances of error. The character should also show as little variation within taxa as possible. If it is not possible to satisfy all these rules at once, it is suggested that ease of observation should take precedence.

These points will be illustrated from the data matrix shown in Fig. 1 and the keys of Figs. 4 and 5. Which characters are easy to observe? This means those which can be assessed with the unaided eye, and which will be observable regardless of the condition of the specimen, i.e. in this example, with or without flowers or fruit, and living or preserved. If we assume that the plant has stem and leaves at least but not necessarily flowers or fruit, characters 5, 7, 9 and 10 could be a first choice. Which of these is best distributed? Character 10 (leaf base shape) is best, with 6 taxa in one state and 7 in the other. Next comes 5 (stem lines) with taxa in the ratio of 4 to 9. Character 7 is a little variable and character 9 is missing in many cases, so dismiss these. Although character 10 is better distributed, its two states are not very clearly distinct, so character 5 will be preferred as a starting character.

In fact, however, the example keys begin with quite a different character, illustrating another principle. If there exist taxa which are strikingly different from the others, it is often convenient to dispose of these first, at or near the beginning of the key. Some authors strongly recommend this procedure as a desirable feature of key making. It may not always be a matter of choice since, if there are taxa which really are very distinct from the others, it is quite likely that they will show states which contrast in characters which are chosen near the beginning of the key. In other words, it may be that the only convenient way to make a key involves distinguishing 'unusual' taxa first. In the example there is one highly distinct species, *E. brunnescens*, which has a creeping and rooting stem, unlike all the rest, so this is keyed out first. There is no necessity to do this in this instance, as other keys could be made which leave *E. brunnescens* to appear later.

Examination of Fig. 1 will show that *E. brunnescens* is the only species to have the character states 'prostrate' for character 1, and 'stem rooting at nodes', as opposed to 'not rooting' for character 6. These characters are diagnostic for this species as they are unique to it, and are alone sufficient to identify it. There is in fact another diagnostic character for this species, flower position (12). This is not so useful as the others, as flowers might not be available, but it is put in as an *auxiliary character*. It is another piece of information which could help to confirm the other distinctions made in lead 1 of the key.

So far we have chosen leads 1 and 2. One taxon has been eliminated. Considering the first part of lead 2, 'stem more or less terete', we can temporarily dismiss any taxa which do not have this character state, leaving just four taxa, i.e. numbers 1 to 3 and 10. Which characters can now be used? Only those which still show some variation among this group of taxa. Looking only at columns 1 to 3 and 10 in Fig. 1, one finds that only characters 3, 4, 8, 10, 13, 14 and 15 show any differences between taxa, characters 1, 6, 12 and 18 are the same for each, characters 9, 11, 16, 17 are incomplete or missing, and characters 2 and 7 are rather variable. Of the first group, characters 3, 4 and 10 divide the four taxa neatly into two pairs, and are leaf and stem characters. Character 10 has states which are not very distinct, so that is dismissed again. Both characters 3 and 4 require the use of a hand lens, but are clear cut. However, character 3 ('direction of stem hairs') is a conditional character depending on whether the stem has simple hairs, and in fact three out of four of the taxa can at times be 'glabrous' (without simple hairs). Hence character 4 ('presence of glandular hairs') is chosen for lead 3.

Now that lead 3 has been established, auxiliary characters can be sought to add to it. If there are other characters whose states are distributed in the *same way* as the states of character 4, these could be used to confirm the division. This is not quite the same situation as in lead 1, where the auxiliary characters were also diagnostic characters. Characters 3, 8, 10, 13 to 15 are the possibles but only 3 has the right distribution. Hence lead 3 could be written:

> 3   Stem with glandular hairs and spreading simple hairs.
>   Stem without glandular hairs, with appressed simple hairs.

However, simple hairs may not be present, so it would be better to put:

> 3   Stem with glandular hairs; simple hairs, if present, spreading.
>   Stem without glandular hairs; simple hairs, if present, appressed.

If information was available about the frequency of plants with or without simple hairs, the following might be better:

> 3   Stem with glandular hairs, usually with spreading simple hairs.
>   Stem without glandular hairs, glabrous or with appressed simple hairs.

This auxiliary character is now logically correct, but less straight-forward than it might have been, since it is now qualified by doubt ('usually') and with an alternative ('or'). Since characters in keys ought to be easy to understand, it can be argued that this auxiliary character is more complicated than useful. In the example keys, it has in fact been left out for this reason. Characters of habitat and distribution can be used as auxiliary characters, but should not be used elsewhere in keys, on account of the possibility that an organism can always be found in a new habitat or locality.

Continuing the process of building the key, the first part of lead 3 concerns just two taxa, *parviflorum* and *hirsutum*. Hence lead 4 is concerned just with listing differences between the two taxa, which 'key out' here, and terminates the branching of this part of the key. It may not necessarily be a good idea to use all the distinguish-ing and diagnostic characters at this point, but only those which are most striking, for if many characters are available, the result may just be an essay describing a taxon, which is better placed elsewhere.

The construction of the remainder of the key continues in the same manner until all taxa have been separated. It is usual practice to put the shorter leads of branches first when writing out a key. For example, the first part of branch 2 has two other branches depending on it and following it (3 and 4), whereas the second part has seven (6 to 12). The sole purpose of this is to reduce the distance which the eye has to travel down the page when moving to the later leads of the same question. Although the example keys are strictly *dichotomous* (with all leads in pairs), there is no strict logical necessity for this. Dichotomous keys are strongly preferred by many authors, especially in the yoked form, because of the possibility that if there are more than two leads in a question, the later ones could be overlooked. A key which includes *polytomous* questions (with more than two leads) may nonetheless be a more sensible choice in some cases, where characters are best ex-pressed with many states. With *Epilobium*, a key to flowering plants could start with flower colour, but this has three states, so that the first question might say:

1	Petals white to pale pink	*roseum*
	Petals rose	2
	Petals pink	3

This could be expressed just as well as a dichotomy like this:

1	Petals white to pale pink	*roseum*
	Petals pink or rose	2

or as:

1   Petals white to pale pink   *roseum*
     Petals coloured otherwise   2

which is not so good, because the use of 'otherwise' does not make clear what the alternatives are. If, for example, the specimen had 'petals blue', it would not be obvious that something was wrong. Another way of avoiding a polytomous character is to split it into a sequence of binary characters, e.g. 'petals pink or not pink', 'white to pale or not', 'rose or not', which is clumsy, and not recommended.

Another aspect of key construction is that of *taxon weighting*. This is the situation where one or more taxa are so frequently met with that it is desired to use as few questions as possible before these taxa are recognized. To put this another way, the length of the path through the key to such taxa is to be as short as possible, in order to save effort whenever a specimen of such a taxon is being identified. This is not the same situation as that for the 'unusual' taxon, because the common taxa may not have any particularly diagnostic characters. In order to write a key with weighted taxa, one must attempt to distinguish those taxa first, and so characters will be chosen in order to separate these taxa in particular rather than all taxa in general. Hence there may be some compromise with the general principles for key construction.

One may normally expect that a key can be used to separate every taxon concerned from every other. Indeed, if not, this could be a warning signal that the classification of the group is inadequate. In special cases, where a key is intended, say, for immature or preserved specimens only, there may be taxa which are not distinguishable with available characters, and then a *partial key* can be made up. The ultimate leads of such a key may give the names of more than one taxon, instead of one only. When such a key is in use, one may find that it is only possible to reach a short list of alternative identifications instead of a definitive one.

Although in the example discussed above, variable and missing characters were given very little weight and were hardly used, there are times when they have to be used for lack of any alternative. The procedure then is to allow such taxon to key out in more than one place. If there was an *Epilobium* species which could have leaves with or without stalks which belonged under the second part of lead 2 ('stem with raised lines'), then it would have to appear under both the leads labelled 6. Also, if these two character states are known to be easy to confuse in this taxon, one could deliberately put in both cases to reduce the chances of error. Likewise, if the state of this character was simply not known, and as there are two possible states, it could be treated in the same way. However, a character state which is inapplicable must not be treated like this, since it can never occur anyhow. It is recommended that missing or variable character states are used only sparingly, and are best avoided if possible. This is because they increase the total length of the key, especially if used at the beginning. To put this another way, they have less information content than constant or complete characters and tend therefore to be less useful.

If the organisms for which a key is wanted are polymorphic, there ought to be a separate key for each form, e.g. for differing male and female organisms, or for the different stages in the life cycle of insects. Although, when the specimens are collected, there might be a considerable population so that different forms of what are presumably the same taxon are found in association, the key writer cannot assume that all the alternative forms are available and must not use characters from different forms in the same key.

Finally, a key must be tested after it has been constructed. The usual way to do this is to try it out with an adequate selection of correctly identified additional specimens, other than those used in the compilation. It is best for some person other than the author to do this to avoid bias. Errors which may be encountered are errors of fact (in the data matrix), or errors made in the actual construction of the key. It may also happen that the characters are not weighted as sensibly as they might be, or that commonly occurring variations in the taxa have not been allowed for. It may not be wise to alter the key in order to allow for every observed variation or aberration since, if these are included, the essential distinctions between taxa may be obscured or destroyed. The writing of a key often proceeds along with the revision of a taxonomic group, and if descriptions of taxa are also being written, these must agree exactly with the key. Some authors put facts in the key which are not repeated in the corresponding descriptive text, so that information has to be sought in several places. This practice is not recommended.

2.1.3 Other forms

The two previous sections described common forms of the diagnostic key, but other varieties exist. So far, questions have been distinguished by numbering them, but some authors prefer to use letters. For longer keys this requires the use of repeated letters, e.g. 'AA', or of other alphabets, and the result can be untidy.

In order to save space on the printed page, one can abandon the use of a new line for every fresh lead, and use a prominent type case, e.g. '**bold**', for the question numbers and taxon names, so that it is still possible to trace a path through the compressed text of the key. A 'solid' key in this form is not very attractive.

Keys with drawings incorporated into the leads are very present-able, but perhaps more expensive to publish. An example is given in Fig. 6 (Chinery 1973). Such keys are most helpful for beginners and for teaching purposes. The choice of an organism to illustrate presents some difficulties, especially at the higher levels, because, however typical the illustration, actual specimens are likely to differ consider-ably from it at times.

A more special form of key is one where contrasting leads of the same question are given different, instead of the same, numbers. The number of the next contrasting lead is given in brackets. Leads, starting from the first are labelled 1a, 1b, etc., until a taxon is keyed out, when the lead number is changed to 2, and so on. This results in a key which also indexes the taxa by the lead numbers. It does not seem to have become widely popular.

Synoptic keys are sometimes published, which look just like conven-tional keys, but are keys intended to present a classification, rather than to identify actual taxa. The classification may well have been simplified or idealized, so such a key should not be used to make identifications.

Yet another type of key exists which begins to look more like a tabular key (p. 33), but its total effect is actually just like that of a diagnostic key, since all specimens must necessarily agree with all the characters used. Rypka (1971) calls it a *truth table*. An example is given in Fig. 7, which is based in the *Epilobium* data of Fig. 1. A certain number of good characters are chosen to begin the key, in Table A. In this case there are three, each with two states, making in all $2^3 = 8$ possible combinations of states. The states are numbered 0 and 1, and when the combinations are written out they become the same as binary numbers (as used in digital computers). For example, the combination 101, regarded as a binary number, is the same as the decimal number 5. This is called the *truth value*, and acts as an index number for the various combinations. In order to identify a specimen one must establish the character state combination which applies, and look for the taxon numbers given at the right. If there is just one taxon, as number 2 for value 5, then the identification is complete. If there is no taxon, as in rows 0 and 5 of Table B, then there must be an error. The taxon numbers given in brackets in Table B really belong in other tables. If there are several taxa, as in rows 2 and 3 of Table A, additional tables are required to separate them (B and C). There is no compul-sion to use only binary characters as, for example, Table B includes character 13 which has 3 states. It is not essential to number states from zero, or to use numbers at all. Some keys have letter codes for the states instead. A key of this type is best suited for situations where characters are best evaluated in groups, or as batches of tests.

Fig. 6. Example of illustrated key (after Chinery).

Key to the Orders of European Insects

1. Insects winged   2
   Insects wingless or with vestigial wings   28

2. One pair of wings   3
   Two pairs of wings   7

3. Body grasshopper-like, with enlarged   Orthoptera
   hind legs and pronotum extending
   back over abdomen

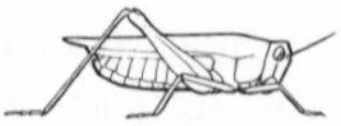

   Insect not like this   4

4. Abdomen with 'tails'   5
   Abdomen without 'tails'   6

5. Insects <5mm long, with relatively   Hemiptera
   long antennae: wing with only one
   forked vein

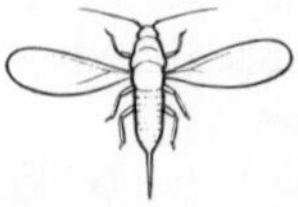

   Larger insects with short   Ephemeroptera
   antennae and many wing
   veins: tails relatively long

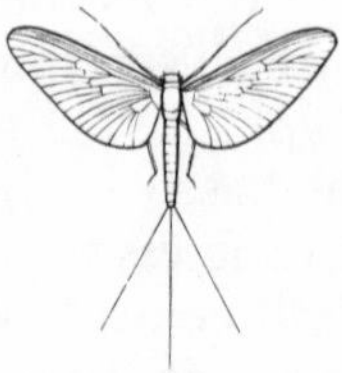

6. Front wings forming club-shaped   Strepsiptera
   halteres

   Hind wings forming halteres   Diptera
   (may be hidden)

7. Front wings hard or leathery   8
   All wings membranous   13

8. Front wings horny except for   Hemiptera
   membranous tip

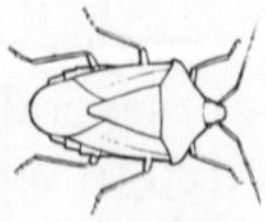

   Front wings of uniform texture throughout   9

A ALL TAXA

Truth value	Character no.			Taxa
	4	5	10	
0	0	0	0	10
1	0	0	1	3
2	0	1	0	Group 1
3	0	1	1	Group 2
4	1	0	0	1
5	1	0	1	2
6	1	1	0	5
7	1	1	1	6

Character	4	no – 0
		yes – 1
	5	terete – 0
		lined – 1
	10	cuneate – 0
		rounded – 1

B TAXA GROUP 1

Truth value	Character no.		Taxa
	7	13	
0	0	0	(5,6,10,13)
1	0	1	4
2	0	2	8
3	1	0	11
4	1	1	7
5	1	2	(1)

Character 7	no – 0
	yes – 1
Character 13	< 6mm – 0
	6–10mm – 1
	> 10mm – 2

C TAXA GROUP 2

Truth value	Character	Taxa
	1	
0	0	9
1	1	12
2	2	13

Character 1:	erect – 0
	decumbent – 1
	prostrate – 2

Fig. 7. Truth table for British species of *Epilobium*.

2.1.4 Theory

The most useful results are set out here with a minimum of mathematics, and without proof. For more details see Osborne (1963).

Firstly, there is a fixed relation between the number of taxa (T) and the number of questions (Q) in a dichotomous key, namely

$$Q = T - 1$$

In the keys of Figs. 4 and 5 the number of questions Q is 12, and the number of taxa T is 13, which illustrates this theorem. Although the number of ways to construct a key for a given set of taxa is very large, the number of questions required will always be the same. If a key is not strictly dichotomous, i.e. some questions contain more than two leads, then Q may be reduced, and the theorem is no longer exact. If taxa are variable, and have to be keyed out in more than one place in a key, then T represents the effective number of taxa including these additions.

There is a connection between the number of taxa which key out at various levels of a key, and the numbers of the levels. For example, the key in Fig. 3 has six levels, and the number of taxa t_i which key out at level i are:

$$i \quad 1 \quad 2 \quad 3 \quad 4 \quad 5 \quad 6$$

$$t_i \quad 0 \quad 1 \quad 0 \quad 5 \quad 5 \quad 2$$

The total number of levels (L) is 6

The relation is:

$$\sum_{i=1}^{L} 2^{-i} t_i = 1$$

In this example the left hand side is

$$\frac{0.1}{2} + \frac{2.1}{4} + \frac{0.1}{8} + \frac{5.1}{16} + \frac{5.1}{32} + \frac{2.1}{64}$$

which adds up to 1.

The theory of errors in keys is generally rather complex (Osborne 1963), but the following simple argument is quite illuminating. Assume that the probability that each question is answered correctly is p, and that all questions are equally likely to be answered correctly. Then the probability that both the first questions will be answered correctly is p^2, and so for a key with L levels, the probability of getting a correct identification is approximately p^L, because the number of questions to be answered before identifying a taxon is, on average, about the same as the number of levels. Suppose $p = 0.9$, i.e. there is a 90% chance of answering each question correctly (or a 10% chance of getting it wrong). With $L = 6$, as in the example key (Figs. 3–5), $p^L = (0.9)^6$, which is roughly 0.5. In other words, even with such a short key as this, the chances of getting a right answer are only even, or fifty-fifty! In reality p is not the same for all questions, and must be a good deal better than 90% if keys are going to work. Also, the actual value of p will not be known, but instead one may only know that some characters are more reliable than others. If an error occurs in using a key, and a question is answered wrongly, the wrong lead is followed, and the key user continues in the wrong part of the key. The nearer the beginning of the key this happens, the worse the consequences may be. Hence, other things being equal, reliable characters are best used first in the key, and the less reliable ones later, so as to lessen the consequences of possible errors.

It is also clear that a key will be more reliable if the average path length for the taxa, which is the average number of questions to be answered in order to key out a specimen, is as small as possible. Two extreme cases are shown in Fig. 8. The average path length in case (a) is 3, and for case (b) is

$$\tfrac{1}{8}(1 + 2 + 3 + 4 + 5 + 6 + 7 + 7) = 4\tfrac{3}{8}$$

Intuitively one can see that the 'best' key of case (a) is arrived at by constructing the key so that each at each question the taxa are divided equally into each lead. It may not be possible to do this exactly, owing to the way in which the states of actual characters are distributed, and

Fig. 8. Tree diagrams for the best and worst cases of average path length in keys for 8 taxa.

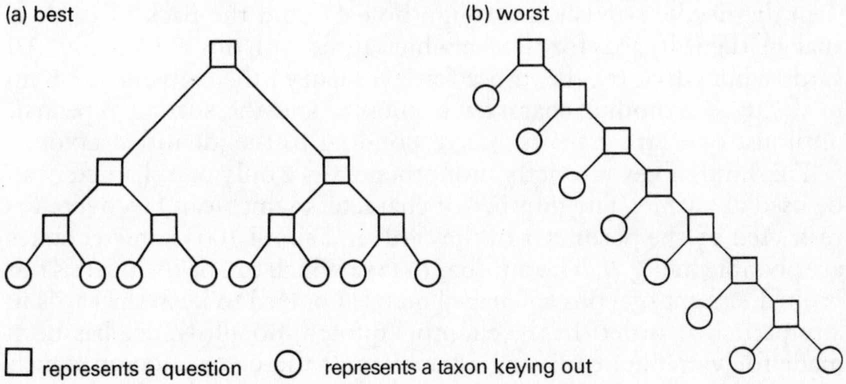

(a) best   (b) worst

☐ represents a question   ◯ represents a taxon keying out

the number of taxa will not usually be a power of 2 as it was in the above example. Hence one may say that a better key is likely to be produced if each question divides the taxa into two sets which are as equal as possible, and that this policy tends to give the shortest average path length. This is not always the case, however, because counter examples can be constructed. This is, therefore, not a mathematical theorem but a useful rule of thumb.

There is a relation between the minimum number of levels L and the number of taxa T. This can be seen in the idealized key, case (a) in Fig. 8. If there are two taxa, one question will be enough to separate them. For four, three questions will be needed, one at level one and two at level two. In the example, 8 taxa are keyed out in 3 levels. One can see that, by continuing this process, L levels can take at most 2^L taxa, so

$$2^L = T$$
or, $$L = \log_2 T$$

Obviously, this is the extreme case, so in general

$$L \geqslant \log_2 T$$

2.1.5 Multi-access keys on punched cards

These occur in two distinct forms, the *edge-punched key*, (Fig. 9) and the *body-punched key* or *feature card* (Fig. 10).

In an edge-punched key, there is one card to each taxon. The states of characters are represented by holes around the margin, and the area of the card may be ruled off so that character state names can be written in, or else these can be specially printed in advance, as shown in Fig. 9, which was designed by P. E. S. Whalley. If a taxon shows a certain character state, the corresponding hole is clipped out, e.g. for 'eyes smooth', number 17 in Fig. 9. Conversely, one could choose to punch out a hole if the character does not occur. In order to use the key, the cards are sorted by a needle. Suppose a specimen has 'eyes smooth', then the needle is pushed through hole 17, and the pack of cards is shaken, then all cards for all taxa which agree with this will fall out. All cards which disagree (i.e. those for 'eyes hairy', the converse) remain in the pack. Another character is chosen, and the sorting repeated, until just one card remains, corresponding to the identified taxon.

This kind of key is strictly monothetic, since only one character can be used at a time. The number of characters which can be covered is restricted by the perimeter of the card, and about 100 character states are possible in Fig. 9. The number of taxa which can be included is not limited, except by convenience. There is no need to keep the cards in any particular order. In the example quoted, no allowance has been made for variable or missing characters. If there was a taxon which

Fig. 9. Edge-punched card from key to *Lepidoptera*.

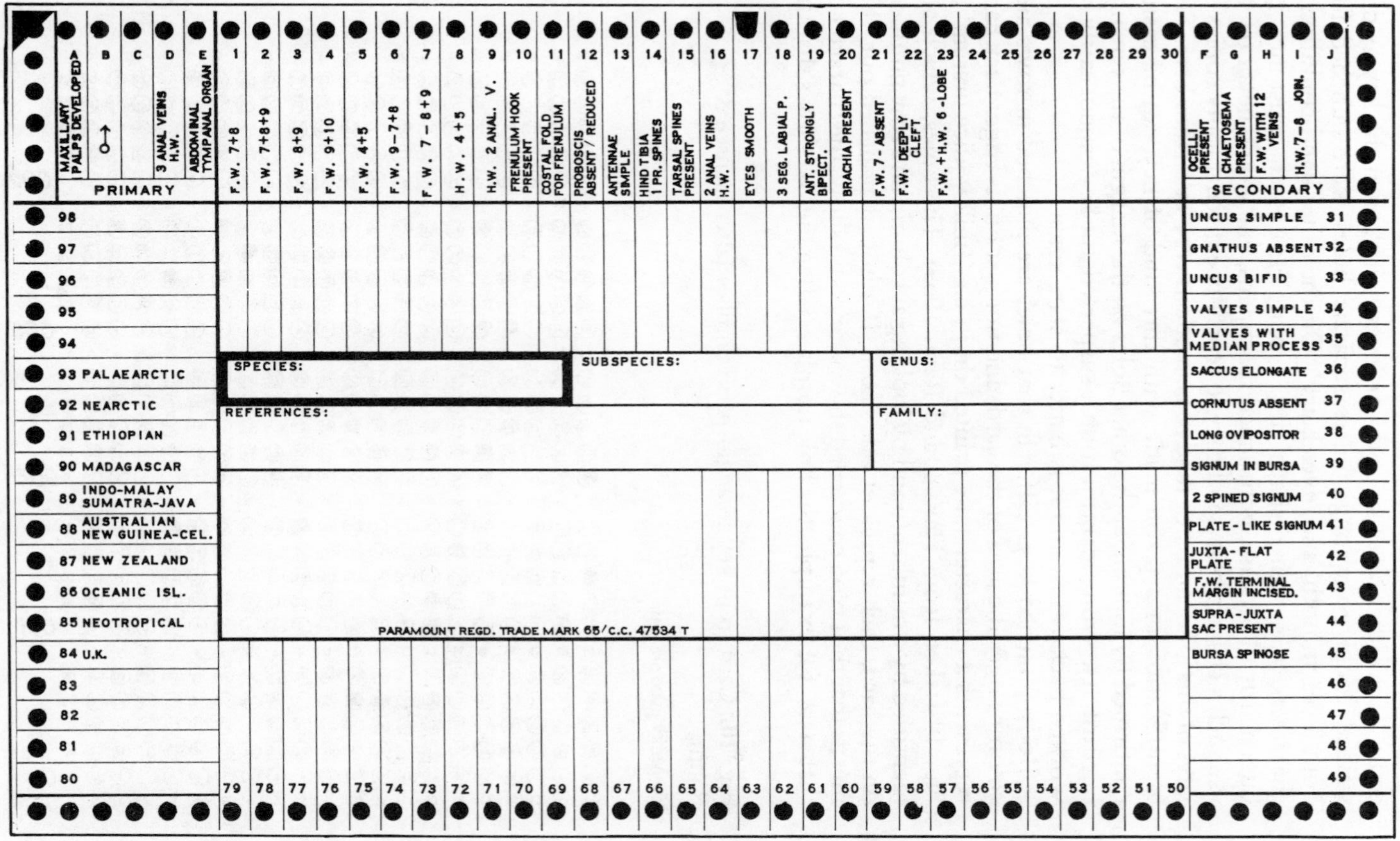

could have either 'eyes smooth' or 'eyes hairy', then the key as described could go wrong, unless separate holes were punched for both 'eyes smooth' and 'eyes hairy'. On the other hand conditional characters, e.g. 'wing shape' for a wingless insect, must not be punched at all because, with actual specimens, one would never try to use such a character! It is not easy to reproduce mechanically copies of keys of this kind, and they tend to be made in very small numbers. The identification of a taxon made with this kind of key can be easily confirmed by looking at the other holes on the margin of the remaining card. If a wrong identification is suspected, the key cannot easily be worked backwards, and it is better to start again.

Body-punched keys differ in that holes are punched in rows and columns over the area of the card, and not just on its edge. Each card represents a character state printed on it, e.g. 'milky juice present' in Fig. 10 (taken from Hansen and Rahn 1969). The holes, which are identified by a number found by looking at column and row numbers on the card, each represent one taxon. If a hole is punched out, this means that the taxon concerned shows, or could show, the character in that state; for example, the hole number 320 in Fig. 10 shows that the

Fig. 10. Card from key to angiosperm families (after Hansen & Rahn).

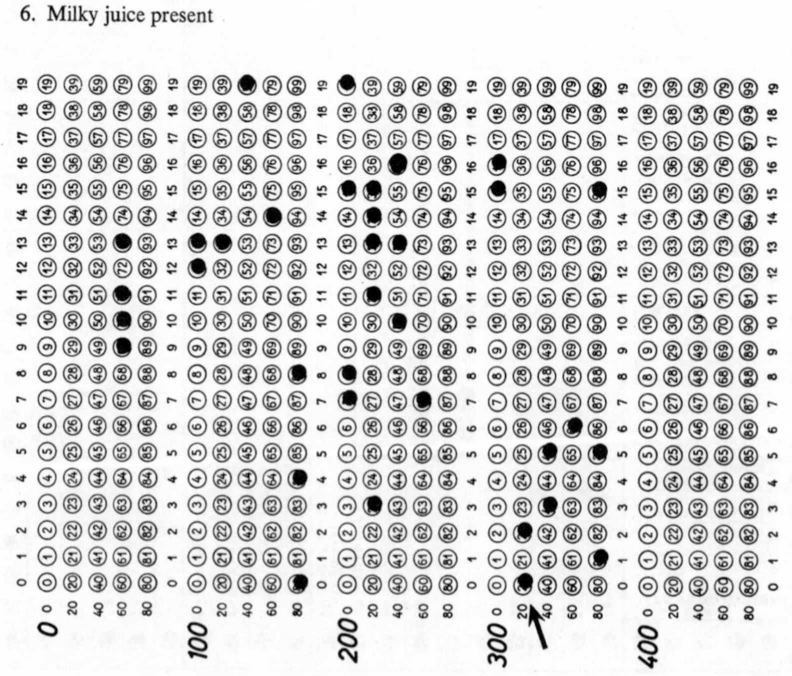

6. Milky juice present

family *Compositae* has character state number 6. There is a separate list of numbers for taxa (families in this case). The key is used by taking out the cards for character states shown by the specimen, and overlapping them precisely. Any taxa which agree with all the character states will show a hole right through when the cards are held up to the light, or laid on a dark background. When only one hole shows, then look up the number of the hole to find the name of the taxon which has been identified.

This key is more or less monothetic, but there is the possibility that if a hole is only blocked by one or two cards, this will still be visible, so that taxa which do not quite agree can still be recognized. This depends on the thickness and nature of the material of which the cards are made. If semi-transparent, coloured or grey plastic is used for such a key, it could permit taxa to be identified correctly even when there are errors, provided the material is neither too pale nor too dark. Any number of characters can be used, but the number of taxa is restricted by the area of the card. In the example (Fig. 10) this is 500 and is not really prohibitive. It is convenient to keep the cards in a fixed sequence so that characters can be found when wanted, but after an identification has been carried out the cards have to be put back in the right place. Sometimes cards are pinned together at a corner and rotated in a fan in order to overlay them, but it can then be difficult to get them properly in line unless there is only quite a small number of them. Variable characters are dealt with by punching a hole in each card which corresponds to a possible state. As before, inapplicable characters must be left out of the key completely. Body-punched keys are more readily reproduced in quantity than edge-punched cards, but still require special equipment, unless computer cards are used (but see p. 53). The identification of a taxon can be confirmed by pulling out further cards for supplementary characters and adding them to the cards already withdrawn. It is easy to back-track from a wrong identification by removing some of the cards previously selected. This is especially useful when there is no hole common to all the cards which have been withdrawn!

Both kinds of punched card key operate by step-by-step elimination, as does the diagnostic key. However any selection of characters may be used, and in any order, and this is why they are *multi-access* keys (sometimes called *polyclaves*). This is an important advantage for incomplete or fragmentary material. It is also possible to stop short of identifying a specimen down to a single taxon if the material does not provide enough information, and to obtain a short list of taxon names for a result. In other words, punched card keys can easily be used as partial keys. They are, however, just as likely to fail if errors are made in observing characters as are diagnostic keys. In general, punched card keys are not very convenient to publish, and hence have not very often been met with in the past.

Multi-access keys are sometimes published by writing out lists of taxa for each character state, without using punched cards. For example, using Fig. 1, the taxa which have, or could have, 'leaves sessile' are numbers 1, 2, 7, 9, 10, 11 and 13. The key is then used by working through a sequence of chosen character states, and rejecting any taxon whose number does not occur in all the lists. A key published in this form can of course be used as an instruction kit for making a set of punched cards for oneself.

Although punched card keys are the most popular form of multi-access key, mechanical equivalents also exist. An instance of this is the 'information sorter' (Olds 1970) which uses transparent plastic sheets instead of cards, one for each character, mounted in a cabinet with internal lighting. The sheets are ruled in three squares for each taxon (microbes), mounted behind a plate with square holes in it, and they can slide to a left, right or central position, corresponding to 'absent', 'present' or 'not used'. The holes in the plate correspond to the central square on the slide when the slide is in its central position. The squares are tinted red for character states which do not agree with a taxon and a green filter is usually switched in to make a contrast. There is a button for sliding the sheet for each character into place, and taxa which agree show as a green square on the front of the machine.

2.2 Identification by comparison

The method of comparison, or matching, is straightforward in principle. The unknown specimen is first examined and its characters are observed or investigated by experiment. It is then compared with all of the taxa of an appropriate group and identified with whichever of these agrees exactly, or is the most similar. The comparison of the specimen with the reference taxa may involve counting the number of characters which agree, or calculating some measure of the agreement (a *similarity coefficient*) based on character matches and mis-matches. The result of the comparison may also be a short list of taxa rather than just one, from which a subjective choice must be made.

The simplest form of this method is that of direct comparison, where one takes the specimen and looks through a book of illustrations, or photographs, or through a collection of preserved specimens, until one finds a taxon which 'looks the same'. No objective assessment of similarity or identity may be used at all, or at any rate, not consciously. This is what many people would naturally do to identify or 'name' something, in the sense described on p. 1; they take a popular handbook on natural history, e.g. on birds, and just thumb through the pages in order to 'find a picture of it'. However, it may be difficult to make this visual assessment without considerable practice if the taxa are numerous and very much alike. Also, if there are no reference collections or illustrations, but only a Flora, handbook or monograph with purely written descriptions, the comparisons are much harder to make without some visual aid.

Comparison methods have the advantage that they do not demand any particular characters from the specimen and can be used successfully with incomplete material. The characters are used all together, polythetically and as the outcome does not depend on the correct observation of every character, some errors can be tolerated. However, the results become more reliable as more characters are used, and so a detailed and perhaps tedious examination of the specimen may be called for. Also, if there are many taxa to consider, there may be much labour involved in comparing the specimen with every taxon.

2.2.1 Tabular methods

These may take a variety of different forms, but the basic procedure can be illustrated from Fig. 1 for *Epilobium*. The figure shows a rectangular table giving the data matrix for this genus. Species are in columns, and characters in rows. The entries in the table are the states of the characters in abbreviated form, e.g. by character 1, 'stem habit', E stands for 'erect'. A new specimen of *Epilobium* is examined and each of the available characters is written down the edge of a sheet of paper, using the same abbreviations as in the table. Move the paper across the table, column by column, noting the number of disagreements for each. If only one species is found which entirely agrees, that will be the identification; otherwise, one could take the species with the fewest disagreements as the answer. If there are several which disagree to the same extent, a choice between them will have to be made in some other way, e.g. by comparing with a collection of preserved specimens. Notice that the disagreements were counted, not the agreements. This is because the number of characters available for comparison varies from species to species because of missing or conditional characters. For example, there are only four species where 'leaf shine', character 11, will contribute, and if the stem of the specimen is hairy (character 2), one cannot compare it with 'habit of hairs' (character 3) in three species. Alternatively, the proportion of agreements could be used, instead of the total, but this extra calculation for each taxon becomes a little tedious to carry out with pencil and paper.

An improvement on the tabular method is the *lateral key* (Fig.11) devised by Sinker (1975). There is a column for every state of every character, not just for each character as in Fig. 1. The characters are

Fig. 11. Part of lateral key to common British grasses (after Sinker).

Species	1	2	3	4	5	6	7	8	9	10	11	12	13	14	15	16	17	18
False Brome (*Brachypodium sylvaticum*)	■	○	○	■	○	■	○	○	■	○	○	■	○	○	■	○	■	○
Italian Ryegrass (*Lolium multiflorum*)	■	○	○	■	○	○	○	○	■	⊙	○	■	○	○	○	○	■	○
Perennial Ryegrass (*L. perenne*)	■	○	○	■	○	○	○	■	■	○	○	■	○	○	■	○	■	○
Common Couch (*Agropyron repens*)	■	○	○	■	○	⊙	○	■	■	○	○	■	■	⊙	■	○	○	■
Bearded Couch (*A. caninum*)	■	○	○	■	○	■	○	○	■	○	○	■	■	⊙	■	○	■	○
Crested Dogstail (*Cynosurus cristatus*)	○	■	○	*	*	■	*	*	■	○	■	■	■	■	■	○	■	○
Wall Barley (*Hordeum murinum*)	○	■	○	*	*	■	*	*	■	○	■	○	■	■	○	○	■	○
Sweet Vernal-grass (*Anthoxanthum odoratum*)	○	■	○	○	■	■	○	○	○	○	■	■	■	⊙	■	○	■	○
Meadow Foxtail (*Alopecurus pratensis*)	○	■	○	○	■	○	■	○	○	○	■	⊙	■	■	○	○	■	○
Timothy (*Phleum pratense*)	○	■	○	○	■	■	■	○	○	○	■	■	■	■	○	○	■	○
Yorkshire-fog (*Holcus lanatus*)	○	■	○	○	■	○	■	○	○	○	■	■	■	■	○	○	■	○
Creeping Softgrass (*H. mollis*)	○	○	■	○	■	○	■	■	○	○	■	■	■	■	■	■	○	■
Tufted Hairgrass (*Deschcmpsia cespitosa*)	○	■	■	○	○	○	■	⊙	○	○	■	■	■	○	■	○	■	○

mm | 10+ | 6–10 | 2–6

■ YES   ⊙ USUALLY NO   ○ NO   * INAPPLICABLE

marked off by vertical lines. Each position in the table is filled with a symbol. The user is provided with, or can make for himself, a strip of card with tabs which can be folded over. The tabs are the same width as the columns. The specimen is examined, and the tabs are folded back for each character state which is observed. The strip is then slid down the table. The taxon which is the correct one will display only black squares (for 'yes') and perhaps some of the signs for 'variable' when the strip is covering the right row of the table. The other symbols, for 'no' and 'inapplicable' will remain covered by tabs which have not been turned down. The advantage of this version is that it is easy to assess visually whether the specimen agrees with a taxon without needing to count the number of agreements.

Fig. 12. The logoscope; example for diagnosis of rabies (disease 12) (after Nash).

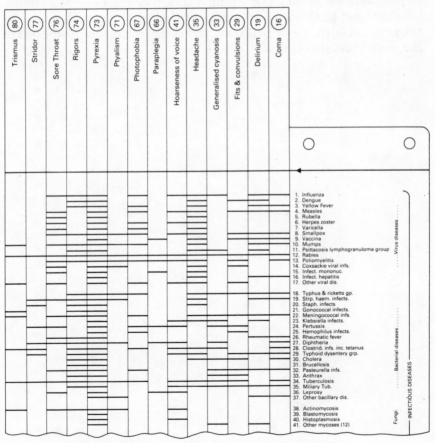

2.2.2 Mechanical methods

Mechanical equivalents of tables have also been tried. One version consists of a transparent plastic slide on which the description of the specimen is written (Cowan and Steel 1960). Underneath this a board is placed on which the table has been drawn. The slide is moved over the table until a row of the table beneath is seen to agree. An improved version of this has been called the 'logoscope' (Nash 1960) and resembles a slide rule (Fig. 12). The table of data is divided into strips, one for each character (symptom or sign), and the user picks out only those strips for the characters which he wishes to use. The strips are marked with a line for a taxon (disease) if the character is present, and left blank otherwise. The strips are put into the bed of the device, and a slide can be moved over them to help detect a continuous or nearly continuous line, representing agreement or partial agreement with a taxon. The user is recommended to use as many characters as possible. This device could be used either as a multi-access key, using step-by-step elimination (as in a body-punched card key), or else as a tabular identification method. Some errors in character observation can be tolerated, since they give rise to a nearly complete line across the bed of the device and this is easily seen.

3 Automatic Identification Methods

The methods which may or must involve calculations are discussed here. The calculations are often somewhat lengthy or complex, so that the use of a computer is usual but not essential. Desktop or pocket calculators, or pencil and paper could also be used. Some of the methods involve the computer indirectly (diagnostic and punched card keys), while others use it directly (matching and on-line). The related problem of character set minimization is discussed, as well as techniques requiring special equipment other than computers. Finally, a critical comparison of the methods is made, so that one can see how to choose a method appropriate to a particular problem.

3.1 Computational background

3.1.1 Advantages and disadvantages of computing

The advantages of using computers include the great rapidity with which complex decisions and calculations can be carried out, and indeed many computations are now carried out which were never practical to attempt before. It is also very straightforward to make minor changes and corrections and to repeat previous work. Consequently the rate at which revisions can be made is greater, and it is easier to keep up to date. Once a computer program has been thoroughly tested and the data properly checked, there is a very high degree of certainty that the results will be strictly accurate. It is difficult and tedious to verify manually constructed identification aids to the same extent. Computers can be used to produce taxonomic data of high quality in regard to factual completeness and consistency, but this is perhaps a mixed blessing, because the taxonomist may have to make a greater effort in order to achieve this quality.

One disadvantage of computer usage, as with any other novel technique, is the initial learning effort required. A number of minute details have to be correctly set out before a computer will accept a program and its data and give correct results. These details tend to vary between different computers and installations, which can be confusing. The cost of using computers is quite easily measured and may be high, but it is difficult to compare this with the cost of traditional identification techniques, since this has scarcely, if ever been measured. This is particularly so when new techniques have no previous counterpart, as is the case with the on-line identification programs.

3.1.2 Data Format

Before a computer can make use of taxonomic data for identification, the data must be coded or represented in some way; this is called the *data format*. The data to be represented include the descriptions of the characters in terms of the states which occur, and the names of the taxa, together with other details such as character and taxon weights if required.

The way in which the data are represented is essentially a compromise between what is convenient for human beings, and what is convenient for computation. To humans, descriptions of taxa are most easily understood as pictures, and also in writing. Computers manipulate facts most readily if they are in numerical form, because in this form programs are easier to write and the machines run faster. While it is possible to have computers read descriptions of taxa in the form in which they usually appear in textbooks i.e. as a simplified form of ordinary language, it is cheaper and more convenient to code the facts as numbers, to some extent at least.

The example of a data format discussed below is not in any sense a standard or an ideal format. Other formats could be used more or less equally as well in order to impart the same information. Two types of format are frequently encountered, *fixed* and *free*. Fixed format means that numbers and words are expected to occur in a regular sequence. For example, a common form of input is the 80-column punched card, and it is customary to punch numbers on these with equal spacing, e.g. one to every space of five columns, and in such a way that each number is *right-adjusted*, i.e. placed at the right-hand end of its space. In fixed format, any number which is not in its correct place may be mis-read. With free format, on the other hand, the items of data need only to be separated by spaces or some other punctuation marks and can be spread out as desired, without regard to the position on a line. Fixed format input is often used with the FORTRAN programming language, commonly used for scientific purposes, and will be used in the examples here. Free format is more convenient for the user, but takes a little more programming.

The example is based on data for moths (*Lepidoptera*). Characters and the states of characters are given numbers. The states are numbered separately for each character. For the character 'proboscis presence' there could be two states, 'absent' and 'present'. This could appear as:

```
1   2PROBOSCIS/PRESENCE
1     ABSENT
2     PRESENT
```

In the first line, '1' is going to be the number of the character, and '2' is the number of states it has. On the next line, the first state (absent) is numbered '1' and on the last line the second state (present) is numbered '2'. We have here also stated that 'absent' and 'present' are the only states applicable to this character. Capital (upper case) letters have been used since these are standard for computers, whereas small (lower case) letters are not always allowed. The character is named as PROBOSCIS/PRESENCE in case there are other characters of the proboscis, such as length, and the oblique (/) shows the computer that PRESENCE is a comment, which it may ignore. In this way the computer can combine the character name with a state name to get a readable phrase, such as PROBOSCIS ABSENT. If the character name had been, say, PRESENCE OF PROBOSCIS, the phrase PRESENCE OF PROBOSCIS PRESENT might occur, which is ugly. Alternatively, one could store two complete phrases, PROBOSCIS ABSENT and PROBOSCIS PRESENT, but this would take up more of the computer store. Some characters are more awkward to express, such as 'male' versus 'female' for the moths. The character name is 'sex', but one would not normally write e.g. 'sex male'. The name of the character can be left out, while just keeping a comment e.g.

```
5   2/SEX
1     MALE
2     FEMALE
```

So far only qualitative characters have been dealt with. Suppose we want to describe 'length of wing', and this varies from 5 to 20 mm. This could be approximated as a qualitative character e.g.

```
10  3WING/LENGTH
1     UP TO 10 MM
2     10 TO 15 MM
3     OVER 15 MM
```

The awkwardness of this will be felt when taxa just overlap two of these states. A taxon with wings 8–12 mm long would have to go in both states 1 and 2, and appear as if it were 0–15 mm! Some improvement can be made by increasing the number of states, e.g. 'up to 6 mm', '6–8 mm', '8–10 mm', '10–12 mm' etc., but such fine divisions may have little meaning.

The ideal way to describe a quantitative character would be to state its statistical distribution with parameters, e.g. normal, with mean and standard deviation. However, the labour required to establish this kind of data for biological identification usually makes it impractical, and measurements are usually quoted as a range, e.g. '5–15 cm', or less often as '(2) 5–15 (20) cm', meaning that the extremes (2–5, or 15–20 cm) are less often encountered. This could be given a little more precision by recording the frequency of different measurements, and giving the ends of the range as percentiles, say at 10% and 90%. This would mean that 20% of specimens measure less than 5 or more than 15 cm. The simplest solution is to give the ends of the range and a unit of measurement, if any. There are two kinds of quantitative character: discrete (e.g. number of spines, having only integers as states) and continuous (e.g. length of wing, having 'real' numbers). Data to define such characters could appear as:

16   SPINES/NUMBER

where '16' is a character number or, similarly, with units:

23   WING/LENGTH/CM

These characters would need to be identified in some way as quantitative, and as discrete or continuous. This could be done by segregating them under separate headings within the data.

Now that characters have been defined, one can set out the descriptions of taxa. This is conveniently done in a rectangular table, or matrix, which could be done with taxa in either rows or columns. If the matrix is given in taxon order, adding or removing taxa is simple but altering characters is more inconvenient. The reverse is true if the data are put in character order, hence neither order is ideal. Suppose each row represents a taxon, then each column represents a character. The number in row 7, column 5 would represent character 5 of taxon 7. This will be adequate as long as no taxon varies and shows only one state for each character. This variation will be especially marked for higher taxa, such as genera or families. The problem is to find a way of putting several numbers into the table where there is only space for one, a difficulty caused by the use of fixed format.

As an example of this situation, consider the description of forewing venation in a family of moths. The veins are numbered, and '+' means, roughly speaking, 'joined to'. There could be six different situations, expressible in the data as:

```
3   6FOREWING VEINS/BRANCHING
1     7 + 8
2     7 + 8 + 9
3     8 + 9 + 10
4     8 + 9
5     9 + 10
6     7 + 8 WITH 9 + 10
```

The states are recorded in the data not as the numbers 1 to 6 but as a series of multiples of 2, namely:

state number 1 2 3 4 5 6
represented as 1 2 4 8 16 32

Hence state 3 ('8 + 9 + 10') would appear as 4, and '8 + 9' as 8. If more than one state occurred, such as '7 + 8' and '8 + 9' and '9 + 10' (states 1, 4 and 5), this would be represented as 1 + 8 + 16 = 25. The number 25 can only be broken down into one combination of the numbers 1, 2, 4 etc., and uniquely represents this situation. This device is not very convenient for those preparing the data, but it is a suitable way to store and process variable characters in computer store. This is because digital computer arithmetic is mostly based on the number 2, rather than the more familiar arithmetic of decimal numbers (base 10). It then becomes a simple matter to test by computer whether a taxon shows a particular state of a variable character or not, and to decide whether two taxa with various states of a given character are distinct or not.

There are two special situations where states of characters are unknown (missing) or inapplicable. There is a difference between these two situations. If a character is unknown, more research or fieldwork can be undertaken to discover it. If a character is inapplicable, its occurrence is logically impossible. In the format, unknown is expressed as '0', and inapplicable as '−1'. To illustrate this, consider character 3 above (vein branching). If this pattern is '7 + 8 + 9' (state 2), there are two possibilities, expressed in character 4, viz:

4 2FOREWINGS/7 + 8 + 9
1 9 FROM 7 + 8
2 7 FROM 8 + 9

So for a taxon with veins 7 to 9 joined in the second way, character 3 = 2 and character 4 = 2. If character 3 had state 2, but character 4 had just been overlooked, then character 4 = 0. If however, character 3 has state 1 (veins 7 and 8 joined, but not 9), then character 4 = −1, because it means nothing. If character 3 = 3 and 4 = 1, this would mean venation could be '7 + 8' or '7 + 8 + 9', and if and only if the latter, it is '9 from 7 + 8'.

The computer will need to know that there is a condition that says 'you cannot have character 4 if character 3 = 1, etc.', and this will also have to be put in the data. For each such rule one may put three integers, e.g.:

3 1 4

which means 'if 3 equals 1 you can't have 4'.

So far only the tabulation of qualitative characters has been mentioned. There will be two numbers for each taxon where a quantitative character is concerned instead of just one, and they would have to be tabulated separately from the qualitative characters to avoid confusion. If some taxa show only one state of a discrete quantitative character, e.g. number of spines = 2, while others have a range, this can be quoted as the number with a blank or as the same number twice.

Weights for characters and for taxa, if required, can appear as a list of pairs of numbers; number of character or taxon and the weight itself. Characters may be weighted in such a way that the weights are actually used in the computations, e.g. one might state that one character was 10 times better than another, or else just signify the order in which characters are to be used. The weights of taxa usually signify the frequencies in which they are expected to occur.

Once the data have been completely scored, the computer can help to check them for correctness. If fixed format is used, this will first mean that all items which have a fixed position must occur where they are expected. A data format such as that outlined above has a certain amount of redundant information in it. This is done deliberately in order to assist checking. For example, one could manage without numbering the characters and their states, provided that one was certain that they were in the correct order. The computer can check that this is so if the characters and states are numbered. The number of states for each character is given, and hence the maximum possible score for each character can be found. For a binary character this is $1 + 2 = 3$, for a 3-state character it is $1 + 2 + 4 = 7$, and in general for an m-state character, it is $(2^m - 1)$. Any entry in the matrix which is greater than this is clearly wrong. Similarly, the code for inapplicable (-1) should only occur if the character concerned is known to depend on another character, and if that controlling character has the appropriate state(s).

The checking procedures just described cannot detect errors of fact, where the description of the taxon given to the computer is not the correct one. The computer can help indirectly by converting the numerical descriptions back into words and printing these, so that the stored data can be verified against the original and errors can be found and corrected.

Apart from getting the data self-consistent and factually accurate, there are other ways in which they can be refined. It is usual practice only to include characters which vary within the group of taxa concerned. The computer can check that this is in fact so, and if not, state which the redundant characters are. One can then make alterations so that the characters are no longer redundant, or they can be weighted so that they are ignored, or can be taken out altogether. Usually all of the taxa are supposed to be different from each other. The computer can compare all of them with one another in pairs, and report on any pairs which are not distinct. It often happens that two taxa are not identical, but have no characters whose states do not overlap, which is what is meant by 'not distinct'. The cure is to redefine the taxa slightly or add new and diagnostic characters. On the other hand, one may wish to continue and produce partial keys instead of complete ones and be content if some taxa cannot be separated.

3.1.3 Estimates of the usefulness of characters

This section describes two ways of calculating which characters are better for constructing keys than others. Questions of the convenience, cost, and reliability of using the characters are ignored; only their ability to separate taxa one from another are considered here.

One such measure is the *separation number*, or the *separation coefficient*. As proposed originally by Gyllenberg (1963) for a character with two states, the separation $S = n_1 n_2$, where n_1 is the number of taxa showing state 1 and n_2 is the number of taxa showing state 2. If none of the taxa is variable, then $n_1 + n_2 = N$, the number of taxa. Hence the equation

$$S = n_1 (N - n_1)$$

is a function of n_1 (or n_2) only, and is greatest when $n_1 = N/2$. This is the situation where the character divides the available taxa into two equal groups which, as was explained above (p. 27), is the ideal for constructing dichotomous keys.

This could be generalized to characters with more than two states, but it is then inconvenient to compare binary with multi-state characters because S ranges over different values, and it needs to be normalized by dividing by some number so that S varies only between 0 and 1 for any kind of character.

Another way of looking at this is to define a *separation coefficient*, s. This is obtained by looking at all possible pairs of taxa, to see whether the character separates each pair or not.

$$s = \frac{\text{No. of pairs of taxa separated}}{\text{Total no. of pairs}}$$

This is a property of the character, relative only to a given set of taxa.

For N taxa, the number of possible different pairs is the number of combinations of two out of N, or $^N C_2$, which is $N(N-1)/2$. For a binary character, the maximum value of s is approximately 1/2, for a 3-state character 2/3, and so on, so that for an m-state character, the maximum is roughly $(m-1)/m$. This approximation is close, and is more nearly exact for a large number of taxa. Hence s varies between zero and $(m-1)/m$, which is less than 1. One could make all the values of s lie in the same range by defining s' as $ms/(m-1)$, so that s' lies between 0 and 1 always. However, one could argue that the amount of useful information in a character ought to increase with m, so it may be best to use s and not s'.

The coefficient s can equally well be defined for two or more characters at once. As the pairs of taxa are considered, one can examine whether any one of the relevant characters, or some combination of them distinguishes the two taxa. For two characters, a and b, s_{ab} will usually be larger than s_a or s_b alone, unless a and b have exactly the same distribution of states over taxa, or if either s_a or s_b is zero.

Another measure for the usefulness of characters in key construction is the *information statistic*. This was originally worked out for problems of errors in the transmission of signals in telecommunication. There is a mathematical derivation of it (e.g. Feinstein 1958) which is remarkable for the amount that can be deduced from very simple assumptions. It is not given here but some background is provided.

Suppose we consider a binary character. The proportion p_1 of state 1 for a group or taxa is n_1/N. Likewise, for state 2, $p_2 = n_2/N$, and since $n_1 + n_2 = N$, $p_2 = 1 - p_1$. The 'information' we get from seeing state 1 on a specimen is the effect this fact has on our opinion of what taxon we think we have. If p_1 was 1 (i.e. all taxa show character state 1 only), then on seeing state 1 we should have gained no information at all. If $I(p_1)$ stands for the 'information' gained from observing state 1, then the average 'information' H we get by observing this character is:

$$H = p_1 I(p_1) + p_2 I(p_2)$$

Evidently, if we observe the same set of characters but in a different order, the final gain of information ought to be the same regardless of the sequence. Expressed in the above rather general terms, that is about enough to decide what the function H must be, and it is:

$$H = -p_1 \log p_1 - p_2 \log p_2$$

or, for a character with m states;

$$H = -\sum_{i=1}^{m} p_i \log p_i.$$

The minus sign is put in because $p_i < 1$, and all values of $\log p_i$ are negative, so that H then turns out positive.

If $p_i = 0$, $p_i \log p_i = 0 \log 0$. Logarithm of 0 is minus infinity, but $0 \log 0$ is defined as 0 for convenience. If $p_i = 1$, $p_i \log p_i = 0$ again. Hence H is zero for $p = 0$ or 1, i.e. for constant characters, which is what one would expect. One can also show that H is a maximum for any m when all p_i are the same and equal to $1/m$. In other words, H is largest if all states are equally distributed. The maximum value is log m. If $m = 2$, we can confine H to the range 0 to 1 by using logarithms to the base 2. For larger m, the maximum of H would be log m. If one wanted to confine H always to the range 0 to 1, then logarithms to base m can be used for characters with m states, because log $_m m = 1$. This is in effect just the same as multiplying H by a normalizing constant.

3.2 Key-constructing programs

A number of computer programs exist which can construct diag-nostic keys, e.g. Dallwitz (1974), Hall (1973), Morse (1974), Pankhurst (1971), Payne (1975) and Ross (1975). The following account does not describe the workings of any of these in particular, but gives a general account of the methods and options. This is because development in this field is very rapid and improvements are being made all the time, so that any statement of what a particular program could or could not do would be unreliable.

3.2.1 Program logic

The general sequence of events in constructing a key in a computer is now described. To begin with the data matrix and its associated information is read into store. What happens after this point is broadly similar to the procedure for constructing keys manually (p. 18). At the start all taxa and characters are potentially available, apart from some taxa or characters which may be weighted in some special way in order to show that they are not to be used at all. Note is kept throughout of which taxa are being considered and which characters are applicable to them. Even at the beginning some characters may be redundant. This can be detected by the fact that they show the same state or states throughout or have all the states missing. Special consid-erations apply to dependent characters. Since such a character cannot be used before its controlling character (e.g. 'wing presence' must come before 'wing length') no dependent character is allowed to begin with.

Suppose for the moment that all characters and taxa have equal weight; the next task is to choose a character from among those available. This is done by calculating some figure of merit which measures the suitability of each character for forming a question in the key. The details of this are given below (p. 47), but when this figure has been calculated for every character, the character which scores the best is the one chosen. It is then possible to decide how many leads this key branch will have by counting the number of different states or combinations of states which occur, and to make a list of which taxa show each of these. The program could also consider more than one character at a time, as two or more could be taken together. The chances of finding neat ways of branching a key with several characters in a question decrease rapidly with the number of characters involved, while the computing time (and cost) increases rapidly. Hence it is wise to control the search for multiple character leads in some way, and this can be done by putting an upper limit to the number of characters allowed. The construction of the key branch is not quite complete, because there may be useful auxiliary characters. To find these, use the lists of taxa which go in each lead of the branch, and consider whether there are any characters which vary in exact correlation with the principal character(s). Such characters will have a state or states which is the same within the taxa in each lead but which is different between at least some of the different leads. If any auxiliary characters are found, they must be remembered along with the primary character(s). This is now the moment to sort the leads into an order of increasing size and to store away the details of the key branch which has just been created in computer memory.

Subsequent questions and leads are generated in much the same way in a suitable sequence, but with a reduced set of taxa and characters as appropriate. A suitable sequence of leads could be that of a parallel key or of a yoked key (p. 11). This can also be regarded as a process of crossing out some rows and columns of the data matrix. The character(s) which are used as primary characters in a branch cannot be used again in derived branches further down. On the other hand it may be useful to get the program to use the same characters repeatedly in different parts of the key so far as possible, so that the total number of different characters required is reduced (see Payne 1975). If there is a definite cost associated with characters, e.g. as with tests on bacteria, this can be an advantage. The use of character weights will also tend to reduce the total number of characters used by favouring those which have high weights. The characters which are now available may not all be useful since, among the reduced set of taxa, some of them may be temporarily redundant. This procedure is repeated until every lead terminates with a single taxon, or there are no more characters which separate taxa.

If characters have different weights, there are several ways to take account of this. One is to group the characters according to weight and only consider those characters which are available which belong to the group of the highest occurring weight. This will ensure that characters are used in descending order of weight, providing that the distribution of states among taxa permits them to be used at all. Another way to use character weights is to include them in the figure of merit discussed below. A special case is a weight, for example a negative weight, which indicates that a character is not to be used at all.

3.2.2 The figure of merit

The figure of merit, also called a *separation function*, is devised in order to decide between different possible ways of branching a key. The figure is *heuristic*, which means that it is justified only by the fact that it works, and not necessarily by theory. It represents a compromise between the various qualities that a 'good' key question should have and is a combination of various factors. The figure of merit can be arranged to have the highest value for the 'best' question, so that the scores for the various characters or combinations of characters are searched for the largest score. It can just as well be arranged to give the smallest score for the 'best' branch, and that is how it will be presented here.

(a) The information given by a character   For this purpose, the separation coefficient, s, or the information statistic, I, can be used. These both increase for better characters, and can be made to suit by subtracting from the maximum value. Since both can be arranged to vary between 0 and 1, one could use $(1-s)$ or $(1-I)$ which will then be smallest for the best character. Both can be adapted to cover several characters in combination and can be used on any kind of character. The separation coefficient is perhaps simpler to calculate and has a more obvious interpretation.

The above is easiest to apply when all taxa show only one state for a character. If some taxa show variable states for a character, the separation coefficient is still straightforward to apply, because even when there is variation, the question, 'Does this character distinguish this pair of taxa?' can still be answered. The information statistic can be used in this case if the proportion of each state occurring for a variable character within each taxon is known (Payne 1975).

(b) Dichotomies   If the number of different character states or combinations of states is K, the question will have K leads. K equals two (dichotomous branching) is usually preferred. Any function of K which increases rapidly for K greater than two would do. One could take a power, e.g. $(K-2)^2$ or $(K-2)^3$, or take an exponential function, e.g. $\exp(K-2)$. Such functions would discourage anything other than dichotomous branches.

(c) Equal numbers of taxa in leads   The best arrangement is to have the same number if taxa in every lead. If there are K leads, let the numbers of taxa be n_1, n_2, up to n_K, so that the total is N.

$$\text{i.e.} \quad \sum_{i=1}^{K} n_i = N$$

If all these were equal, then for any i, $n_i = N/K$.

Usually the n_i are not all the same so that

$$\left(n_i - \frac{N}{K}\right)$$

is the difference between the ideal and the actual. In order to remove the effect of a different number of taxa at different stages of the key, it would be better to consider

$$\left(1 - \frac{N}{n_i K}\right)$$

This is either positive or negative, so it could be squared to make it always positive or else the minus sign could just be removed (signified by vertical lines), viz:

$$\left(1 - \frac{N}{n_i K}\right)^2 \quad \text{or} \quad \left| 1 - \frac{N}{n_i K} \right|$$

This could be added together for all leads, e.g.

$$\sum_{i=1}^{K} \left(1 - \frac{N}{n_i K}\right)^2$$

This figure would tend to be larger for larger values of K, and if this effect is not wanted, one could divide by K. In this way we have found a function which will be zero for the ideal and increases for any other arrangement.

If taxa have been given different weights, it is here that this can take effect. This can most easily be done by treating a weighted taxon of weight w as if it were represented w times. If w is 10 for one taxon, for example, then a key branch with this taxon on its own in one lead would exactly balance 10 other taxa each with weight 1 in another lead. Hence the heavily weighted taxon can be keyed out separately at an earlier stage. The above formula is altered by replacing n_i by n'_i where n'_i is the sum of the weights of the taxa in lead i instead of just the number of taxa. Likewise N becomes N' where the weights of all taxa are added and not just their total number taken.

(d) Missing character states   It is preferable not to use characters whose states are missing from the definitions of the taxa, because any taxon whose state is unknown for a character to be used in a key branch has to be put in all the leads in case someone observes the character later. This increases the size of the key and contributes nothing to the identification of the taxon. Sometimes, e.g. with fossils, characters of taxa are often missing and such characters are hard to avoid. In this

case, it may be better not to try and make a key but to use a different identification method instead (see p. 78).

Suppose we have a binary character with some missing states. If the number of taxa with state 1 is n_1, with state 2 is n_2, and with n_0 missing, then

$$n_0 + n_1 + n_2 = N$$

The numbers of taxa, effectively, with states 1 and 2 will be

$$n'_1 = n_1 + n_0$$

$$\text{and } n'_2 = n_2 + n_0$$

since the missing states could occur equally well as 1 or 2, or both, as far as is known. The new values of n'_1 and n'_2 can be used in the formula under *(c)* for testing equal distributions of taxa as before. To discourage the expansion of missing states, the proportion

$$r = \frac{n_0}{N}$$

could be used in the figure of merit or, for a more severe deterrent, just the value n_0.

(e) Quantitative characters   There is a special situation which occurs with quantitative characters. One could have two quantitative characters which were equally good in all respects as above but which still differ in the degree to which they separate the taxa. For example, suppose there are two characters A (range 0–50) and B (range 20–200). Which would be better for separating taxa X and Y, if the states are as follows?

	A	B
X	0–15	150–200
Y	35–50	20–90

The gap in the values of A is 20 (between 15 and 35) and for B is 60 (150 minus 90). Is B therefore better, because the gap is larger, and thereby making a clearer distinction? Not necessarily, because B varies more. It would be better to find the ratio of the gap to the total range, which for A is $20/50 = 0.4$ and for B is $60/180 = 0.33$. Hence, on this count, A is really the better. This ratio increases for better characters, so to put it in the same manner as the other figures we could use

$$1 - \left(\frac{\text{gap}}{\text{total range}} \right)$$

A character of this kind is said to be *bimodal*, and a measure such as the above is used by Hall (1973). A bimodal character could arise with variable qualitative characters which have many states, but this would be a rare situation.

(f) Character weight   This could be treated as contributing to the figure of merit as a separate item. If so, it would need to be put in a form such as

$$(\text{maximum weight} - \text{weight})$$

$$\text{or} \quad \frac{1}{\text{weight}}, \text{ or exp} (- \text{weight})$$

so that the best character makes the smallest contribution.

Finally these figures can be combined into one figure of merit in various ways. They could just be added together, or they could be multiplied by a positive constant before being added together, in order to alter the relative importance given to different aspects of 'goodness' of characters. The figures could also be multiplied together or some could be added and others multiplied into the final figure. One arrangement which would be particularly sensible would be to add together figures (a) to (e) and then multiply by a factor for the character weight. A comparison of some different figures of merit for key construction is given by Gower and Payne (1975).

3.2.3 Programming details

A number of more special details are covered here, including some points of interest to computer programmers.

(a) Partial keys   The usual way to recognize that a key branch has been completed is when all taxa are distinct and the final leads each relate to one taxon only. If in fact all the taxa were not distinct from each other, this could be regarded as an error in the data matrix. Occasionally an incomplete or partial key may actually be desired, in which case the branching of the key may terminate with more than one taxon in a lead. The reason for stopping may be that there are no characters left by which taxa could be separated.

(b) Duplicate taxon names   Usually all the names of taxa in the data base will be different, but occasionally several forms of one taxon will all be described under one name. In this case the branching of the key must cease as soon as all the taxa in a lead have the same name, even if there are several and even if there are still characters available. As an example of this, a data matrix might include species from several genera. If the species all had different names, a key to the species could be made, but if only their generic names were given, a generic key would result.

(c) Unusual taxa   Programs for key construction could be written

which attempt to identify taxa which are 'unusual'. This would require estimating what the description of an average taxon would be, and finding those which are least similar beyond some arbitrary limit. Even so this would not guarantee that the 'unusual' taxa would always have some convenient diagnostic characters for separating them. The easiest course seems to be simply to decide subjectively which the 'unusual' taxa are and to give high weightings to their diagnostic characters, which will have the desired effect. If there are no obvious diagnostic characters, taxon weighting could be used to give these taxa prominence, provided that suitable distributions of character states exist. There does not seem to be any particular need to write program statements specifically to deal with 'unusual' taxa.

(d) Variation of characters within taxa   A method of describing characters in the data matrix which permits any variation to be included within a coding of one integer number has already been described (p. 40). This may have appeared awkward but is conveniently arranged for the internal logic of computers. Integer numbers are stored as a sum of powers of two, and hence a state recorded, as in the previous example, with the number 25 is stored as $16 + 8 + 1$, or

$$1.16 + 1.8 + 0.4 + 0.2 + 1.1$$

which is

$$1.2^4 + 1.2^3 + 0.2^2 + 0.2^1 + 1.2^0$$

This is stored as a *binary* number as 11001, where the 0's and 1's are just the multipliers from in front of the powers of two. Practically every digital computer is provided with two simple instructions, called a *logical sum* and a *logical difference*. The logical sum of two numbers is a pattern of 0's and 1's, such that there is a 1 if both the corresponding positions are 1, and 0 otherwise. This is in fact the answer to the question, 'Have these two taxa any states in common for this character, and if so, which?' Likewise the logical difference answers the question, 'What are the states shown by one or other or both of two taxa for this character?' The use of this internal representation for characters is therefore very convenient for answering questions such as these within a program, regardless of whether the characters vary or not.

(e) Internal storage of keys   While a computer is constructing a key, it has to store the partly completed version in its memory in order to be able to find out which taxa and characters are appropriate to each new branch. A convenient way to do this is to use what is called a *list structure*. That is to say, while the details about the leads of any one question are kept together in consecutive words of the store, called a

block, the different blocks are scattered throughout part of the store which is called a *free storage area*. A record is kept of how large each block is, and where it is. Special subroutines permit the program to 'ask' for blocks of suitable size when they are needed, and to relinquish them later when they are no longer required. The need to 'give up' blocks of store can arise when a question is constructed but is replaced by a better question found subsequently. The list structure also contains *pointers*, which means that a note is kept within each block saying where the next one is. These pointers are like the lines joining the boxes in the tree diagram of a key in Fig. 3. In this context, the 'next' block is the one which corresponds to the next question which would be written on the page when the key is expressed in words. There are therefore two ways of ordering the blocks, according to whether the sequence of a parallel or yoked key is used. This can be described another way by saying that the order for a parallel key corresponds to tracing out the tree of the key (Fig. 3) across the levels from left to right before moving down the levels and that the order for a yoked key is from top to bottom before left to right. A useful way of keeping track of a partly completed key is to use what is called a *stack*. This means keeping a note of which was the last block considered at each level, and a note of the current level number.

While in most key-constructing programs it is convenient to use a list structure for storing the key, it is not essential to do so. Dallwitz (1974) stores the key as a rectangular table which is written on top of the initial data matrix of taxa and characters.

(f) Printing of keys   Some programs print the key in a numerical form which can then be written out by hand to give a finished product. Others go on to print the text of the key. This basically means substituting words for the numerical version of the key, plus laying it out neatly on paper. A parallel key, or else a yoked key, is obtained merely by choosing one of the two alternative ways of following the list structure (see last section). If indentation is required, this is easily found from the level of each key branch in the tree (Fig. 3). If the question is at level L, it must be moved $(L-1)$ spaces to the right. The program will distinguish between leads which end in the names of taxa and those which point to other leads. Punctuation with commas and full stops must be inserted and, if a sentence is too long for the width of a page, it must be broken between two words and continued on the next line. Some computers can only print upper case (capital) letters, but where possible lower case (small) letters can be put in as appropriate. Further sophistication can be obtained by attending to the left to right order of characters in multi-character leads, e.g. for flowering plants it is customary to follow a sequence of vegetative, then floral and finally fruiting characters when writing a description.

However, the principal character must still come first. Information to describe this desired order would then need to be added to the data format, which was not included in the above discussion. It is also possible to improve the phrasing of the sentences by removing repeated words e.g. the phrase

Wings present, wings 5–10 mm long

could be simplified to

Wings present, 5–10 mm long.

3.3 Punched card keys

Punched card keys, their uses, advantages and disadvantages, have already been described (see p. 28). The universal punched card used by computers has 80 columns and 12 rows and is body-punched across its entire area. Hence computer-produced punched card keys are of the body-punched type. Although a computer makes them in the first place, they are used in the hand as before. An example of such a card is shown in Fig. 13, from a key corresponding to Barnett and Pankhurst (1974). The card can be printed in many different designs and colours, but generally has the columns numbered from 1 to 80. In this case the numbers are printed below row zero and at the bottom. The rows are identified as 0 to 9 by printing rows of digits across the card, except for the top two rows which are left blank. The top edge of the card is generally used for printing, and there is a standard coding scheme for combinations of holes punched in columns which correspond to letters, numbers, and punctuation signs. Using the 10 rows and 80 columns, there is space for 800 holes. Each card represents a character

Fig. 13. Card from computer-constructed punched card key.

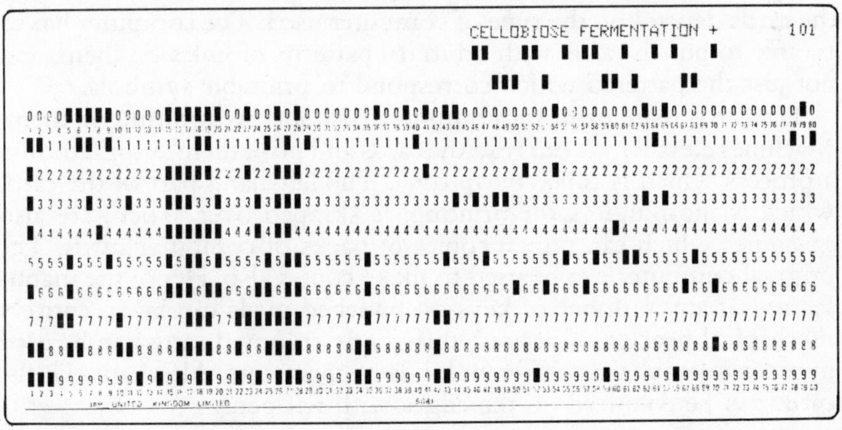

state, so that up to 800 taxa could be put in. In fact, the right-hand end of the card in Fig. 13 has been punched with holes which are codes for the character state and a card number, as printed in the top right margin so in this particular example 44 columns have been used for taxa. The key from which this card comes covers 434 species of yeasts. The taxon which is represented by a punched hole is easily identified by its column and row number. For instance there is a hole in column 11, row 3, so this is species number 113 (*Candida steatolytica*). Because the columns of the standard computer card are numbered from one and not from zero, the list of taxa with their numbers begins with number 11, which is really the first taxon. In the example the holes from column 46 onwards have nothing to do with the taxa; they are only there as a means of putting in the printing at the top right.

Keys can be punched onto computer cards by hand using card punching machines. This is tedious, inflexible, and prone to errors, so it is much better to use a computer to make the cards. Computer programs for making punched card keys have been described, e.g. More (1974), Pankhurst and Aitchison (1975) and Shultz (1975). These are quite straightforward. The data matrix has to be read in and then the computer has to go through each character and each state of each character. It then has to find which taxa show, or could show, each state and arrange to produce a card with holes in the right places. For variable characters several cards will need holes for the taxon concerned. For taxa with missing characters holes ought to be punched in cards for all states of the characters. Inapplicable characters do not need to have any holes punched for their states, since these will never be observed. The computer can also arrange to punch holes on the card for the character state and number. In the example (Fig. 13) the character is number 1 (cellobiose fermentation) and its state is also number 1 (+), and the card has been numbered 101. Similarly, state 3 of character 16 would be numbered 1603. The details of punching the cards depend on the type of computer used. The computer has to be able to punch cards with arbitrary patterns of holes on them, and not just the patterns which correspond to printable symbols.

The cards issue from the computer without any printing on them. Machines exist which can read the cards and print them automatically, a process which is called *interpreting*. The left-hand part of the card, which is meaningless for printing, is skipped over. There are also machines which can punch copies of packs of computer cards. The original computer can be used to make copies also. Hence the manufacture of any number of keys on punched cards can be an entirely mechanical process. If desired, differently coloured cards can be used in order to emphasize different kinds of characters. Alternatively, the cards can be coloured on the edges with felt pens.

Punched card keys have two great advantages: the user can choose any character, and can use the characters in any order. Nevertheless, they are still sensitive to errors for the same reasons as with a diagnostic key. In the past there was no simple and cheap method for creating and reproducing them, which must have restricted their application, but now a large increase in their use may be expected.

3.4 Matching methods

In this section a number of loosely related methods are described, all of which produce a numerical value by which identification is assessed. This value is produced by a comparison of the unknown specimen with all the taxa in the matrix and by selecting those taxa which give the best results. The necessary computations are generally impractical without the use of calculators or computers, which are now an essential part of the identification procedure. These methods require more effort on the part of the user than those described previously, and correspondingly give more precise and more reliable results, since they are all polythetic and do not demand that the data be without errors. There are, broadly speaking, two kinds of methods involving:

1) a measure of likeness between taxa, such as a similarity coefficient or a correlation coefficient. Included here are Euclidean distance methods where, however, what is calculated is distance or dissimilarity.

2) a measure of the probability of identification, such as those based on Bayes' theorem or by the use of discriminant analysis.

3.4.1 Measures of likeness

The similarity coefficient between two taxa is defined as the ratio

$$\frac{\text{no. of characters which agree}}{\text{total number of characters}}$$

The total number of characters means the total number which are known in common between two taxa or specimens, since they may not show the same set of characters. Characters with missing or inapplicable states are left out. Whether or not two characters agree is a simple question if the characters are binary and not variable within the taxa. For multi-state characters which vary, this becomes more complicated. There is a difference according to whether one taxon is being compared with another or whether a single specimen is being compared with a taxon. In the latter case one cannot necessarily expect a single specimen to show all the variation. This depends on whether the variation, if any, occurs within single organisms or between members of a population. As an example, suppose that a plant taxon can have leaves which are round or elliptic, depending on habitat. A specimen has round leaves. This agrees exactly with the taxon, if each individual plant is only expected to have one shape of leaf. If, however, the taxon has round leaves at the base of the stem and elliptic leaves higher up, and the specimen has only round leaves, the agreement is not exact. For the purpose of comparing a specimen with a taxon in a computer program, the simplest course is to ignore the distinction between variation within and between specimens by assuming the former, and to calculate the agreement for a character as a fraction, namely

$$\frac{\text{no. of states shown in common}}{\text{total no. of states shown by both}}$$

This definition lends itself readily to computation because each state may be represented by a 'bit' in a computer word (see p. 51) and the ratio becomes

$$\frac{\text{no. of bits in the 'and' of the two characters}}{\text{no. of bits in the 'inclusive or'}}$$

Another question about agreement of characters concerns the significance of a negative match. If two taxa both lack a character of the presence or absence type, can one say that this is an agreement or should this be ignored? Practice varies on this point, but it seems more reasonable to allow negative matches when the organisms being compared are closely related, as is usually the case in identification by matching.

If desired, weights can be attached to the characters. These could be chosen subjectively to represent convenience or be calculated on the basis of the distribution of states (separation coefficient or information statistic). In order to give weights to the occurrence of rare states of characters, the weight could be calculated according to the particular specimen. This means that the weight given to a character is not fixed in advance, but is made small if the specimen shows a common state and large if it shows a rare one. If the data matrix includes a mixture of characters with different numbers of states, the many-state characters could be given a higher weight since they are each equivalent to several binary characters.

The question of weighting when applied to conditional characters causes difficulty. Suppose there is a presence-absence character which, if present, can have other characters depending on it. For example, consider the leaf of a plant which might have a margin which is entire (without teeth) or serrate (toothed). If teeth are present, they could have properties of shape, relative size, direction, and so on. A specimen with toothed leaves may be able to score agreement with half a dozen characters in all, whereas one with entire leaves will only score with one presence–absence character. This can be regarded as natural weighting, favouring specimens in their agreement with taxa which have a character present. Alternatively, the bias could be removed by giving a high weight to the controlling character for when it matches by being absent. The weight could be the sum of the weights of the dependent characters.

Finally, the similarity coefficient can be written in a more mathematical form, as

$$\frac{\sum_{i} w_i \, A_{ji}}{\sum_{i} w_i}$$

for the similarity between a specimen and taxon j, where w_i is the weight of character i, and A_{ji} is the agreement between the specimen and taxon j for character i. Many other kinds of similarity coefficient have been proposed which could be used (Chapter 4 of Sneath and Sokal 1973).

For completeness, the formula for the calculation of a correlation coefficient is given. If x and y are two quantitative variables, then the correlation between them, given N samples x_i and y_i of each variable is:

$$\frac{\sum_{i=1}^{N} \sum_{j=1}^{N} (x_i - \bar{x})(y_i - \bar{y})}{\sqrt{\sum_{i=1}^{N} (x_i - \bar{x})^2 \cdot \sum_{i=1}^{N} (y_i - \bar{y})^2}}$$

where $\bar{x}$ is the mean of x, and $\bar{y}$ the mean of y.

This is intended for N samples from two variables, whereas if x and y represent the states of characters of two taxa, each with N characters, the formula can still be used to calculate a number, but it will not have any statistical significance. This number can be calculated without difficulty if all the characters are quantitative continuous variables, but is harder to apply if some or all of the characters are qualitative. In the following account of identification by use of the similarity coefficient, one could use a correlation coefficient instead (see Gyllenberg and Niemelä 1975).

Various programs exist for identification by matching with similarity coefficients, e.g. Pankhurst (1975), Ross (1975). The example of output in Fig. 14 is from the former program relating to *Banisia*, a genus of tropical leaf moths (Whalley 1976). The first line of the results is just a title or comment relating to the specimen, which is a male thought to be of species *fenestrifera*. There follows, among other things, a list of taxa which could be identified with the specimen. These are listed in sequence with the similarity as a percentage, the count of characters used and the taxon name. Only a small proportion of all the taxa are listed, namely those which show the highest similarity. It is then the responsibility of the program user to choose the answer from the information provided.

The similarity values may not necessarily be very high percentages. Experience suggests that a level of 60 to 90% is usual, with an error margin of about 5% depending on various causes discussed below. The count of the number of characters used is given because unusually low numbers of characters cause more error in the similarity, and a high similarity with a low character count ought to be treated with suspicion. For example, the second species in the list, INOPTATA, has a low count (25) but quite a high similarity. This comes about because this species is only known from a female, whereas our specimen was male.

The program provides various ways of helping to choose the final identification. As an option 'special' characters can be chosen. These are written out from the second line of Fig. 14 onwards, e.g. UNCUS

Fig. 14. Example of output of program for matching by similarity (data by Whalley).

```
/BANISIA CF. FENESTRIFERA MALE
  SPECIAL CHARACTERS ARE -
     UNCUS/DIVISION
     GNATHUS/PRESENCE

   SEC    SIM.   COUNT              SPECIES

    1     91.6    53    *++         FENESTRIFERA
    2     78.0    25                INOPTATA
    3     71.0    52     ++         FURVA
    4     70.1    49    *++         INTONSA
    5     66.7    48     ++         IDALIALIS

RESEMBLES GROUP    4

SPECIAL TAXA COMPARED
   1     91.6  *                    FENESTRIFERA


REPORT ON TAXON     15
  CHARACTER STATE
     5      1
        DISTINGUISHED TAXA-      1     2     3     4     5     6     7     8     9
                                10    11    12    13    16    17    18    21    26
                                27    28    29    30    31    32    35

    27      1
        DISTINGUISHED TAXA-      2    11    23    25    27    31    32    35
```

DIVISION. The idea is that the specimen will present various obvious or prominent characters which are such that one would have difficulty, subjectively, in accepting an identity for the specimen which did not agree with most or all of these characters. The rows with plus signs before the species names signify which of these special characters agree. Plus stands for agreement and blank for disagreement. The special characters as named in the list from top to bottom correspond to the columns of plus signs from left to right. For example, the first taxon FENESTRIFERA agrees with both special characters but the second species agrees with neither. The more plus signs there are against the name of a species, the more reasonable this species would be as an identity for a specimen. The program also tries to assign the specimen to an intermediate taxon. Suppose the data matrix concerns a family of species, then the genera are intermediate taxa. The program finds the intermediate taxon which, on average, most resembles the specimen. Any species in the list which is a member of this intermediate taxon (e.g. genus) is marked by an asterisk, e.g. FENESTRIFERA, INTONSA, both in the same subgenus. Hence, one may take those names with an asterisk and the highest number of plusses as the most reasonable candidates. Notice that the word 'reasonable' was used, and not 'probable'. There have been no probabilities or statistical calculations involved, although the answer indicated as above might in fact be the most likely one.

If the user of the program has some prior notion of which taxon the specimen belongs to, this can be specified as an option as shown further down under the heading 'SPECIAL TAXA'. The similarity data are printed again, as a precaution in case they did not appear in the list above. There then appears an account of the character states of the specimen which did not agree with the taxon, if any. In the example (Fig. 14), there were two such erroneous characters. Notice that the identification is still successful in spite of the presence of the errors. This information can be used to revise the description of the specimen or the definition of the taxon. There is also a warning that if the character state is added to the description of the taxon, useful distinctions between the taxon and various others may be lost. For example, in Fig. 14, character 5 for species 15 appears highly diagnostic.

While a program of this kind has some tolerance of errors, the results must still be checked by comparison with full descriptions, illustrations or preserved specimens. Ideally the correct taxon will appear at the head of the list, with the highest similarity, with all special characters agreeing and be assigned to the correct intermediate taxon. The margin of difference between the correct taxon and the next depends on the classification and the number of characters used. If in the classification taxa are grouped or clustered closely together, there will be few characters which differ between them. Also, if a large number of characters are used, this will increase the similarity when a high proportion of the characters agree. For example, if two taxa differ only by one character out of 50, they can at best only be separated by 2%. In this example errors could still be tolerated in the description of the specimen, provided that the critical distinguishing character was correct, since this would only reduce the level of similarity but not the difference. On the other hand, there is then the risk that some other taxon will appear more similar to the specimen than the right one, and come before it in the list. For a data matrix to give the best possible performance, it must contain at least one differing character for each pair of taxa, and preferably more. However, in adding a character to increase the difference between pairs of taxa which closely resemble each other, one is decreasing the proportional difference between other pairs which do not differ on this character. Hence, the most effective matrix, without regard to subjective qualities of characters, will be that which has the highest ratio,

$$\frac{\text{average no. of differing characters for pairs of taxa}}{\text{total number of characters}}$$

provided that all pairs of taxa are distinct.

The above argument assumes that the data matrix is both correct and complete. The correctness of the data matrix can be ensured by careful compilation and checking, but the effect of missing characters is important. In the above account missing characters were simply ignored when calculating similarities. The result of this is that if the character had been known and had it agreed, the similarity would be correct but otherwise is in error by being too small. There is no way to overcome this difficulty except by making further observations to complete the original matrix. The benefits of doing this can be very marked (Pankhurst 1975).

The accuracy of the allocation of the specimen to an intermediate level taxon depends on the classification. The intermediate taxa must be genuine clusters, such that the average similarity between taxa within clusters is greater than that between clusters. The best way to ensure this is to carry out a numerical clustering on the same data matrix using the same type of similarity coefficient, and to choose the clusters from the results. Even so, if the taxa do not in fact possess the structure of clusters, e.g. when they exhibit a cline of variation, one cannot expect to be able to identify accurately intermediate-level taxa.

If the characters of the taxa are continuous quantitative, it is reasonable to represent the taxa geometrically as points in a multidimensional space. The states of the N characters are represented by the perpendicular distance of points to each of the N axes. One cannot picture this for *N* larger than 3, and in Fig. 15 only two dimensions are shown. The distance between two points in the diagram represents the dissimilarity between two taxa. The scale of each axis represents the weight of a character, so that if all characters are to have equal weighting, the scale on each axis should be the same, e.g. from zero to one. This method of calculating the dissimilarity between taxa is called the *Euclidean distance* method.

Fig. 15. An illustration of the four different identification states (after Gyllenberg).

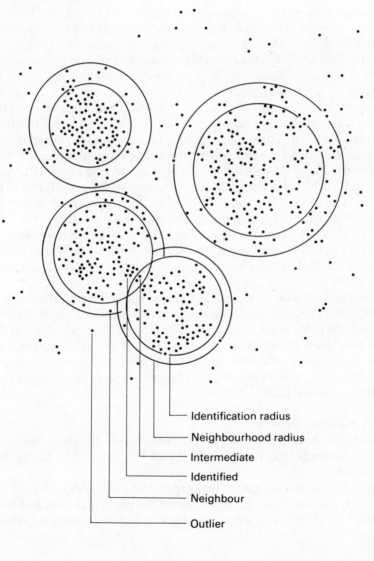

Identification radius

Neighbourhood radius

Intermediate

Identified

Neighbour

Outlier

This approach is particularly suitable for describing what happens to specimens identified by a matching method which do not exactly fit defined taxa (Gyllenberg and Niemelä 1975). In Fig. 15 the variation of taxa is represented by the way in which points are clustered together. The centre of such a cluster is the typical example of the taxon, and can be found by averaging all the characters of all specimens which are believed to belong to it. If we draw a circle (in general a hypersphere) of suitable radius (*identification radius*) about the centre, we can decide to say that any specimen falling inside is correctly identified. We can draw a larger circle with the *neighbourhood radius* to cover specimens which are not near enough to the centre to be correctly identified, but which are still nearer to this taxon than any other and are *neighbours*. A specimen which falls outside either of these circles is an *outlier* and is not identified. We can also have the situation where two taxa are so close that their identification circles overlap, and any specimen which falls in the area common to both is an *intermediate* and cannot be assigned to either. It is clear from this how a computer would be programmed to identify specimens by this method. The choice of the radii is a difficult matter and depends ultimately on the quality of the initial classification. The circles must be large enough to avoid reject-ing too many specimens as outliers but not so large that they overlap and have too many intermediates. An obvious way to choose the radii of the circles is by percentages of the commonest character states or by standard deviations. In practice, however, few data are available on the statistical variation of characters of biological species.

The method of Euclidean distances is an attractive one because of the way it can be pictured geometrically but unfortunately the charac-ters for most biological identification problems are qualitative, not quantitative.

In passing it should be noted that any of the clustering methods of numerical classification could also be used as a means of identification. The unknown specimen would be added to the original data and the results examined to see which cluster the unknown is assigned to. Such an approach would be wasteful because of the extensive compu-tations involved in working out the clusters afresh every time and has not been seriously used.

3.4.2 Probabilistic methods

Under this heading we shall consider Bayes' theorem with the related maximum likelihood method and the method of linear dis-crimination.

Bayes' theorem is a means of making use of knowledge about how often particular states of characters occur for particular taxa and how often the taxa themselves occur, in order to find how likely it is that a specimen with a given pattern of characters belongs to a taxon. As an example of this, suppose that a rare taxon always has a character state which is rarely seen in other commoner taxa. Evidently the occurrence of this character state makes it more likely that we have the rare taxon, but the rarity of the taxon itself makes it less likely. Bayes' theorem enables the probability of this situation to be calculated and compared with that of an unusual character state in a common taxon.

The theorem is given here without proof. It gives the probability that the specimen s is the same as the ith taxon (t_i) out of n

$$P\ (s = t_i) = \frac{P(t_i)\ P\ (s/t_i)}{\sum\limits_{i\ =\ 1}^{n} P(t_i)\ P\ (s/t_i)}$$

where $P(t_i)$ is the probability of taxon i occurring, and $P\ (s/t_i)$ is the probability of s, given t_i, i.e. the probability that t_i has the character states shown by s. The probability $P(t_i)$ acts like a taxon weighting in this formula. $P(s/t_i)$ in turn needs to be calculated from the probabilities $P(c_j/t_i)$ that each character c_j in t_i has the right state, which is the product of all the probabilities of N individual characters multiplied together:

$$P\ (s/t_i)\ =\ \prod\limits_{j\ =\ 1}^{N}\ P\ (c_j/t_i)$$

This is true provided that the characters c_j are independent of each other. If they are not, the proper approach is to find out by experiment what is the probability of occurrence of various combinations of character states which are known to depend on one another, but the number of possibilities to be accounted for may be very large.

The *maximum likelihood* method is closely related to this but gives all taxa the same probability of occurence, i.e. $P(t_i) =$ constant, or equal weight to all taxa so that the formula reduces to

$$P\ (s = t_i)\ =\ \frac{P\ (s/t_i)}{\sum\limits_{i\ =\ 1}^{n} P\ (s/t_i)}$$

This simplification is often used not because the probabilities of occurrence of taxa really are all the same, but merely because they are unknown or variable.

Within the subject of biology applications of these methods are rare outside microbiology and agriculture, and it is worth giving the reasons for this. To begin with much laborious experimental work is needed in order to obtain the probabilities of states of characters. The probabilities of the occurrence of the taxa themselves are usually known in qualitative terms, e.g. common, rare. A more fundamental difficulty concerns the probabilities themselves; are they really probabilities? The definition of probability is the constant ratio of specific events to the number of samples taken. However, there are many situations where individuals of biological species have character states which are known to depend on the environment, according to the availability, for example, of moisture or food. Hence the proportions in which various character states occur within a taxon may be a function of other variables and not constant at all. It is also known that the genetic constitution of different populations of a species can be different, which can also result in varying proportions of character states between individuals in different places. Lastly, it is not always reasonable to suppose that specimens are sampled according to the proportion of variant individuals occurring in the wild (which itself can vary seasonally). The probability that particular species out of a genus occur will usually be very different in different habitats, and biologists themselves can much alter the probabilities by deliberately going to a locality where rare species are known to occur. For reasons such as these, Bayes' theorem or maximum likelihood is usually not applicable unless the sampling and the populations are carefully controlled, as for example with those bacteria which harm human beings or with varieties of crop plants. In situations where taxa overlap, however, a probabilistic method of some kind is a good way to distinguish taxa, and here Bayes' theorem or discriminate analysis are appropriate.

As examples of these methods, there are the programs of Baum and Lefkovitch (1972) for cultivars of oats, and Willcox *et al.* (1973) and Gyllenberg and Niemelä (1975) for bacteria. Some results from the program of Willcox are shown in Fig. 16 and these will be discussed in detail. This is a maximum likelihood method, which operates on a data matrix containing the probabilities of character states. These were established by much careful experimental analysis of numerous examples (strains) of the different species included. Although some of the tests used are known to be (practically) always positive (or negative), the probabilities are never entered as exactly 0 or 1 but are approximated as 0.01 and 0.99 respectively. This is because a zero probability (for either positive or negative results) leads to a zero on the bottom line of the formula for maximum likelihood (p. 63) and the calculation fails. This situation cannot be ignored on the grounds that it has never happened so far because new results or observational

Fig. 16. Computer identification of bacteria by maximum likelihood (after Willcox).

```
FOR: ANYWHERE PHL                              OUR REF:9992/75 RUN W1
DATE: 24/04/75                                    (M619 LAB.1234)

YOUR REF: 123456        ATKINS,THOMAS

COMPUTER IDENTIFICATION BASED ON YOUR RESULTS, 28 TESTS DONE:

NOT IDENTIFIED, FURTHER TESTS SELECTED

YOUR RESULTS USED IN CALCULATION:


MOTILITY 37      -   1      NITRATE         -   5     GLUCONATE        -    1
GROWTH 37        +  99      SIMMONS CITR    +  85     MALONATE         -   40
MACCONKEY        +  99      UREASE          -  50     ONPG             -    1
CATALASE         +  99      GELATIN 1-5     -  25     GLUCOSE PWS      +   99
OXIDASE          -   1      KCN             -  40     GAS GLUCOSE      -    1
H&L OXID         +  99      H2S PAPER       -   1


ADONITOL PWS     -   1      INOSITOL PWS    -   1     MANNITOL PWS     -    1
ARABINOSE PWS    -  55      LACTOSE PWS     -   1     SUCROSE PWS      -    1
DULCITOL PWS     -   1      MALTOSE PWS     -   1


MR 37            -   1      VP 37           -   1     INDOLE           -    1


FURTHER TESTS SELECTED:

     FIRST SET                    SECOND SET
ARGININE         9   1      TYROSINE PIG    8    5
THORNLEYS ARG    9   1      CELLOBIOSE ASS  6   80
INOSITOL ASS     4   1      GLYCEROL ASS    6    1
TREHALOSE ASS    4   1      DULCITOL ASS    4    1
SUCROSE ASS      3   1      FRUCTOSE ASS    3    1
LYSINE           1   1      ETHANOL ASS     2   80
                           PIGMENT          1    1
     SET VALUE =  30/  30        SET VALUE =  30/  30

DETAILS OF CALCULATION:

          GROUP                        SCORE
ACINETOBACTER CALCOACETICUS          .963004
PSEUDOMONAS CEPACIA                  .020087
PSEUDOMONAS PUTIDA                   .006919
PSEUDOMONAS FLUORESCENS              .006362
XANTHOMONAS SPP.(NOT HYACINTHI)      .002391
PSEUDOMONAS FRAGI                    .000933
```

errors can occur. The error introduced by this approximation is never more than one per cent in each probability value and is apparently tolerable.

The program proceeds by calculating the probability for each taxon, and then selects those which score most highly and, in this respect, resembles the similarity program described above. A score for each of the most likely taxa is printed at the bottom of Fig. 16. These scores are actually relative, not absolute, probabilities, since they have been multiplied throughout by a normalizing constant so that they add up to one. The species at the head of the list, *Acinetobacter cal-*

coaceticus, has a score of 0.963, which looks impressive but is not in fact considered adequate for identification. The reason for this is partly that the real probability is not 96% but something different. Also, the criterion for identification used with this program is that the 'right' answer must have a relative probability of at least 99.9%. Put another way, this means that the summed relative scores of all the alternatives must be 0.001 or less, so that the score for the nearest contestant must be 0.001 or less. The actual probability will be different from this score but this will be ignored. Each good character which disagrees will give a factor of 0.01, because of the cut-off at 1%, so this means that a 'right' answer will agree by roughly one to two more characters than the nearest contestant. This could also only mean that the 'right' answer disagrees by a few less characters, most of the rest being totally wrong, so that as a check a list of aberrant results for tests is printed by the program for each species which is positively identified. In the example (Fig. 16), however, no definite identification is reached and so the program computes a list of further tests to be performed under the heading 'FURTHER TESTS SELECTED'. The way in which this is done will be explained in a later section (p. 76). The program also gives, under the heading 'YOUR RESULTS. . . ', the test results originally given (as + or −), together with the matrix entries expressed as percentages for the species with the highest score. It is interesting to estimate the similarity coefficient which corresponds to this situation. If every test result at 1% is regarded as a disagreement, there are 16 of these out of 28, so that the similarity is less than 50% even for the best answer. While identification in this particular program is by relative agreement, Gyllenberg and Niemelä (1975) show how an identification radius and a neighbour radius can be used with the maximum likelihood method.

Missing character state values in the data matrix cause a noticeable loss of accuracy for this method, just as they do for matching (p. 60). The Willcox program gets over this with the following strategy. First, the best possible situation i.e. agreement, probability 0.99, is assumed for each occurrence of a missing value. If no definite identification is achieved with this optimistic assumption, the program just works out a set of further suggested tests. If there is a definite identification, the program repeats with the worst possible assumption, disagreement, for missing values. If after this, identification is still definite, identification is accepted. Otherwise the program works out a set of further tests as before.

A special problem occurs with conditional characters. An example of this for bacteria is growth and motility at 37°C (human body temperature). If a strain does not grow at 37°C, its motility will be negative. The result for motility is not just mostly negative if growth is negative; it is always so. The program has to anticipate and intercept this situation and substitute probabilities of nearly zero instead of zero exactly.

Discriminant analysis is another important statistical method which can be applied to identification, although it is strictly speaking a method for distinguishing between two or more populations. The identification problem is to distinguish one specimen from many taxa, which become analogous to populations if their variability is described. We have to regard the single unknown specimen as an approximation to a population. The basic idea of discriminant analysis is to find a boundary between two populations which is such that on the boundary a specimen is equally likely to belong to either population. In terms of Fig. 15, a line can be drawn somewhere between two identification circles so that character state combinations represented by points on the line are specimens which are equally likely to belong to either taxon or cluster. If we consider just two populations, which are both normally distributed with continuous characters and with same covariance, an estimate of this covariance together with the two population means can be used to calculate a discriminant function of the form

$$ y = \sum_i w_i C_i $$

where C_i is the ith character of the specimen, and w_i is a character weight, derived from the above calculation. If there are two populations, y is negative for one and positive for the other so that an unknown specimen can be assigned to one or the other by calculating y from its characters. This can easily be generalized to many populations (taxa) by calculating a y value for each population and assigning the unknown specimen to the population with the largest y. An explanation of the theory is given in Chapter 8 of Sneath and Sokal (1973).

In fact, most biological identification problems do not involve just continuously varying characters and even if they do, one may well not have normal distributions with equal covariance. In order to cope with discrete qualitative characters, where the normal distribution is not appropriate in any case, the distributions of the characters must be worked out in another way. This can be done by enumerating cases, which is laborious and requires many data. Alternatively, one can assume that the characters are independent and use the maximum likelihood formula to map the distributions, which is also laborious. These practical difficulties detract from what is otherwise a promising method in theory.

3.5 On-line identification programs

The most sophisticated type of identification method available is that which involves interaction with a computer while the identification of a specimen is in progress. The computer is an essential part of the method and makes possible certain techniques which cannot easily be carried out in any other way. The user, who is the biologist wanting to identify a specimen, will have a computer terminal of some sort, with a keyboard for typing instructions and a printer like a typewriter or else a visual display (like a television screen) for getting the answers. A program is said to be interactive, or *on-line*, if the time between typing an instruction or data and getting a reply is short enough to make it seem to the user that the computer is working only for him. In reality the user might be one of many working simultaneously with a large *time-sharing* computer system. On the other hand one might have a small computer which is currently entirely occupied with, or dedicated to, the identification program. The larger desk calculator machines are now getting so sophisticated that they also might be used in much the same way as a small dedicated computer system.

An on-line identification program could be programmed to provide any or all of the identification methods so far described, but because it makes large demands on computing resources, and may be costly to run, it is reasonable to concentrate on those identification methods where interaction permits a step-by-step elimination or a question and answer procedure of some kind. Hence on-line programs, e.g. Boughey *et al.* (1968), Morse (1974) and Pankhurst (1976), tend to work in a manner which is loosely related to the method of multi-access diagnostic keys, as expressed, for example, in the form of a polyclave on punched cards.

Fig. 17 illustrates some aspects of one such program (Pankhurst 1976), using data on *Jurinea*, a genus of thistle-like plants (Kožuharov 1976). The actual man-machine dialogue has been adapted a little in order to make it easier to follow. Any line which was typed by the user begins with an asterisk (absent in reality) while all other lines were typed by the machine. Various comments have also been added, all of which are in lower case. The dialogue is initiated by the user who types various *commands*. In this program a command is a word, such as CHARACTER, followed sometimes by one or more integers. Commands do not have to be typed in full, and in fact all but the first four letters are dispensed with so that, for example, CHAR is enough for CHARACTER.

Example 1 in the figure illustrates a command called BEST, which asks the machine which character is the 'best' one to use next. Ignoring considerations of cost or convenience, the computer decides this by computing the separation number (p. 43) for each character. Other criteria, such as the information statistic, could also be used. The number (16) and name (CAPITULA SHAPE) of the currently best character are printed along with the separation (number of pairs of taxa separated by this character). The user can then ask for the next best character by typing N or give up by typing Q (for quit). After the latter the machine shows that it is ready to accept a fresh command by typing READY. It is now up to the user to pick on one of the characters suggested, whichever is the most convenient. If he chooses character 10 he types CHAR 10 and the machine responds by listing all the available states of this character by number and name. The program is designed this way so that the user does not need to memorize or look up which numbers correspond to which characters and states. It could equally well be designed without this prompting, resulting in a program which is both more economical on computer storage and less convenient to use. The user now responds by typing 2 for state 2, which is what the specimen shows. If he had typed 1 + 2, the machine would interpret this to mean that the character was variable and would remember both states. If he had typed 4, the machine would object and request another number, since state 4 does not exist. The machine now responds by listing the 9 taxa which agree with this character. The user continues by asking for another 'best' character. The answer is computed in terms of the taxa and characters currently available, i.e. the taxa which agree with the specimen as described so far and those characters which have not been used already, and which still vary. By this last is meant that characters which initially showed different states among different taxa may no longer do so once some of the taxa have been eliminated, so they become redundant for the purposes of separating the remaining taxa. How to decide which characters are the 'best' at *any* stage of the identification process is something which can only be achieved in an on-line program. Subsequently, character 3 is selected and scored with state 1, which results in the identification of the specimen as species number 4 with a congratulatory message. Note that if the user had tried to select character 10 again at this second stage, the computer would object because this character has already been used. It would also have objected if the user had tried to use a character which was currently redundant.

Example 2 of Fig. 17 shows another technique which is especially well suited to the on-line program. When asking for the 'best' character in the previous example, the user may not have had any preconceived notion as to which taxon his specimen belonged to. But suppose he had a strong suspicion that it was really a specimen of taxon 4, which characters should he use in order to prove that it is this one and not any other? This is easily answered if the taxon in question

Fig. 17. An on–line identification session (data by Kožuharov 1976)

Example 1

```
• BEST                                          Ask for the best character to begin with
  SEPN  CHARACTER
  91   16 CAPITULA SHAPE                         This is the best
• N                                              but ask for another
  78   10 BASAL LEAVES HAIR ABOVE
• N
  77   24 PAPPUS RELATIVE LENGTH
• N
  66   17 OUTER INVOLUCRAL BRACTS HABIT
• N
  59   23 CORONA OF ACHENE SIZE
• Q                                              Stop asking
  READY
• CHAR 10                                        Choose character 10
  10 BASAL LEAVES HAIR ABOVE
   1  SUBGLABROUS ABOVE                          These are the alternative states
   2  ARACHNOID-TOMENTOSE ABOVE
   3  SETOSE ABOVE
• 2                                              Specimen shows state 2
          9 TAXA REMAIN
  DIFFS NO.  NAME
   0    1   1.J.LINEARIFOLIA                     These are tentative identifications
   0    4   4.J.PINNATA
   0    5   5.J.TANAITICA
   0    6   6.J.ALBICAULIS
   0    7   7.J.KIRGHISORUM
   0   10  10.J.MOLLIS
   0   14  14.J.GLYCACANTHA
   0   15  15.J.HUMILIS
   0   16  16.J.TAYGETEA
  READY
  .....
• BEST                                           Another 'best' character
  SEPN  CHARACTER
  5    16 CAPITULA SHAPE
• N
  4     3 STEM LEAF DISTRIBUTION
• N
  4    23 CORONA OF ACHENE SIZE
• Q
  READY
• CHAR 3
  3   STEM LEAF DISTRIBUTION
   1  LEAFLESS
   2  LEAFY AT BASE
   3  LEAFY THROUGHOUT
• 1
  WELL DONE, ONE TAXON REMAINS                   Identification complete
  DIFFS NO.  NAME
   0    4   4.J.PINNATA
  READY
```

Example 2

```
• DIAG 4                                         I think it is J. PINNATA, but how do I prove it?
  DIFFS NO.  NAME
  120   4    STEM SHRUBBINESS                    Best diagnostic for PINNATA
• N                                              Ask for another
  110  16    CAPITULA SHAPE
• N
  100  24    PAPPUS RELATIVE LENGTH
• Q
  READY
• CHAR 4                                         Describe diagnostic character
  4 STEM SHRUBBINESS
   1 HERBACEOUS
   2 WOODY AT BASE
```

```
• 2
  5 TAXA REMAIN
  DIFFS NO. NAME
  Ø    1   1.J.LINEARIFOLIA
  Ø    3   3.J.TZAR-FERDINANDII
  Ø    4   4.J.PINNATA
  Ø    6   6.J.ALBICAULIS
  Ø    7   7.J.KIRGHISORUM
  READY
```

Example 3

```
• DIFF 3 4                                        What are differences between taxa 3 and 4?
  3 STEM LEAF DISTRIBUTION
  1ST
  3 LEAFY THROUGHOUT
  2ND OR SPECIMEN
  1 LEAFLESS
  2 LEAFY AT BASE
• N                                               Another difference
  5 RHIZOME PRESENCE
  1ST
  2 PRESENT
  2ND OR SPECIMEN
  1 ABSENT
• Q
  READY
• DIFF 1                                          How does taxon 1 differ from specimen?
  3 STEM LEAF DISTRIBUTION                        One difference
  1ST
  3 LEAFY THROUGHOUT
  2ND OR SPECIMEN
  1 LEAFLESS
  2 LEAFY AT BASE
• N
  16 CAPITULA SHAPE                               Another
  1ST
  1 CYLINDRICAL
  2ND OR SPECIMEN
  4 OBCONICAL
• Q
  READY
```

Example 4

```
• LIMI  2                                         Allow up to 2 mistakes in specimen description
  14 TAXA REMAIN
  READY
• TAXA
  DIFFS NO. NAME
  2    1   1.J.LINEARIFOLIA                       Now the number of differences
  2    2   2.J.STOECHADIFOLIA                     between specimen and taxon shows
  1    3   3.J.TZAR-FERDINANDII
  Ø    4   4.J.PINNATA
  2    5   5.J.TANAITICA
  2    6   6.J.ALBICAULIS
  1    7   7.J.KIRGHISORUM
  2   10   10.J.MOLLIS
  2   12   12.J.LEDEBOURII
  2   13   13.J.CONSANGUINEA
  2   14   14.J.GLYCACANTHA
  2   15   15.J.HUMILIS
  2   16   16.J.TAYGETEA
  2   17   17.J.FONTQUERI
  READY
```

has a single diagnostic character state shown by no other taxon. Such characters are often emphasized in Floras, monographs and handbooks. If, however, the specimen is incomplete and lacks the diagnostic character, or if the taxon is fairly nondescript and is distinct only by some non-obvious combination of several characters, an on-line program can solve this problem. The example shows the user typing DIAG 4 (for taxon number 4) and the machine gives a sequence of characters in reply, much as it did for the BEST command. Here, instead of the separation number, a diagnostic score for each character is given under the heading DIFFS. This diagnostic score is worked out as follows.

The score is of one of two different kinds, depending on the situation. If for the given taxon the character in question provides a clear cut difference between taxon and any of the others, the score is the number of such taxa which are different. By 'clear-cut' is meant that either the taxon has only one state for the character or, if it has several states, that none of these partially overlap with the same character on other taxa. This first kind of score is arbitrarily multiplied by a factor of 10, because it is the preferred situation. If on the other hand the character is not clear-cut, but still shows a different range of variation in the given taxon from that which it shows in others, the second score is awarded. This is just the number of taxa which show overlapping character states but which have a different range of variation possible for the taxon, so the second score is not multiplied by any factor because it refers to a potentially much less useful character.

As a result of using the best diagnostic character which is suggested, number 4 with state 2, the computer finds that only 5 taxa remain—a better outcome than that obtained by using BEST, when 9 taxa were left.

Another advantage of an on-line program is in answering some simple questions about taxa which can arise in the process of identification, but which are not always easy to answer with conventional methods. Suppose that the identification method led to one taxon as an answer, whereas one had expected it to lead to another. How then did these two differ? If the two taxa are not closely related it may not be very easy to answer this from textbooks. In example 3 of Fig. 17, the command DIFF 3 4 asks what the differences between taxa numbers 3 and 4. A sequence of differing characters and their states is obtained. A similar question is answered by the next command, DIFF 1. Here the problem is to know how taxon 1 differs from the specimen so far described, because taxon 1 was thought to be a possible identification for the specimen but it has been eliminated (via other commands which are not shown). What is shown here are really just some simple instances of *information retrieval*.

Another facility which is particularly useful in on-line programs is the *variability limit* (Morse 1974). This is a precaution against the effect of erroneous characters in the specimen description. The computer can be told what number of characters is permissible as disagreements. For example, if a limit of 2 is chosen, taxa will not be eliminated from the list of tentative identifications unless they have more than 2 characters in disagreement. This is shown in example 4 of Fig. 17. The user has entered a limit of 2 by typing LIMI 2. The machine responds by merely stating that 14 taxa remain. Previously the machine had positively identified the specimen as taxon 4 with zero differences. In this particular program, a list of current taxa is not produced unless 10 or less taxa are involved to save the typing of too much output. Hence another command, TAXA, is used in order to know which the 14 taxa are. This list is similar to those seen earlier, but there are non–zero numbers under the column headed DIFFS (for number of characters differing). Taxon number 4 has zero differences and is first choice for the identification, but taxa 3 and 7 each have one difference and these could be checked as reasonable alternatives. One could find out where these single differences lay by typing DIFF 3 and DIFF 7.

The effect of a non–zero variability limit is to create a method of identification which lies somewhere in between the sequential elimination of a diagnostic key and the method of matching by similarity coefficients. At the extreme, by setting the limit equal to the number of characters, the number of differences for each taxon will be equivalent to a dissimilarity coefficient. The higher the limit is set, the longer it will take to achieve an identification. If the limit is m, the first m characters described will not eliminate any taxa at all, and thereafter extra characters will be required if the user wants to have zero disagreements for the 'right' answer and more than m disagreements for any other, although one can choose to stop before this point.

Finally, one simple command which is very useful is that for deleting a character which has already been described, but which is later suspected of being wrong. Any previous character can be removed and the effect seen immediately. This is analogous to using a punched card key and taking out a card for a character which had previously been put in. In effect the key is being operated backwards when this is done.

The range of possible commands for an on–line program is very great and only some of the more characteristic ones have been described.

3.6 Character set minimization

The question to be answered here is that of finding a way to make an identification with as few characters as possible. With diagnostic keys the key itself is a tentative answer because of the tendency to divide the taxa into equal groups at each stage, but with the matching methods discussed above it is better to have complete descriptions of specimens. If specimens are going to be described in full, we still want to avoid including characters which are not necessary. It is natural, then, to ask for the smallest set of characters which will separate all taxa, which is the *minimum character set*. This need not be unique as there could be several minimal sets of equal size. In practice, the smallest set might be an impractical combination of characters, so that some other slightly larger set with more convenient characters might be preferred. In addition, the minimum set will very likely include only one differing character to distinguish some of the taxon pairs, and it might be better to ask for at least two distinguishing characters for each of the possible pairs of taxa.

The idea of a *diagnostic character set* or *diagnostic description* is closely related to the minimum character set. This is a set of characters which distinguishes a given taxon from all others. The question is, 'How do I know that I have taxon X, rather than any other?'. If the specimen agrees with every character in the diagnostic description of X, then that is what it must be. Diagnostic descriptions are often given in Floras, monographs or handbooks and are mostly derived subjectively from experience.

There are both exact and approximate methods for finding minimum character sets. The exact method described below (Kautz 1968) finds all sets of characters which distinguish all taxa, from which one can then choose the smallest or most convenient one. As a small example for the sake of illustration, take the first 5 taxa and the first 10 characters of *Epilobium* in Fig. 1. The table given below gives all characters which distinguish all the 10 possible pairs of species. The characters 1 to 10 have been lettered A to J. The rows and columns represent species numbers, so that in row 2, column 3, for example, the characters D(4) and G(7) distinguish species 2 and 3. Evidently there is no need to fill in row 3, column 2, as well because this would only repeat what is already known, nor the diagonals, which represent only the difference between a species and itself.

	1	2	3	4	5
1	–	HIJ	DGHJ	CDEGHI	CEGHI
2		–	DG	DEGJ	EGJ
3			–	EJ	DEJ
4				–	D

There is no restriction in the nature of the characters which can be used, binary or multi-state, discrete or continuous, constant or variable; the states only need to be non–overlapping. There is one catch, however, concerning conditional characters and variation. For example, character C(3) had different states for species 2 and 3, but it should not be put in the table. This is because character B(2) controls character C, and C can only be observed if B has state 'hairy'. Species 2 and 3 are both variable in character B, and so character C is not necessarily observable and cannot reliably be used to distinguish this species pair.

The table of differences can be much simplified. Compare entry 1–3 with 2–3. This shows that species 2 and 3 have characters D and G to distinguish them, and 1 and 3 have D, G, H, and J. Evidently D and G will be enough to keep both pairs separate, and so H and J are not needed under 1–3. We can therefore put DG under 1–3 instead. However, we already have this pair of characters under 2–3, so the entry under 1–3 can be removed altogether. This argument leads to the following rule: if any two entries are such that the characters in the shorter one are all included in the longer one, delete the longer one. This also means that if any entries are identical with others in the table, cancel all but one of them. After these simplifications, the table reduces to:

	1	2	3	4	5
1	–	HIJ	–	–	CEGHI
2		–	–	–	–
3			–	EJ	–
4				–	D

We now have to find all the combinations of characters which will still keep all species apart. Let us begin by taking one character from each of entries 1–2, 3–4 and 4–5 and finding all the different combinations, 6 in all. These are easily seen to be:

HED, IED, JED, HJD, IJD and JJD

The last of these is the same as JD, since 'JJ' only means the same character twice over. Also, JD is included in JED, HJD and IJD, so these latter can be dropped. Putting these in order, we have:

DJ, DEH, DEI

Now combine these sets each in turn with the remaining one in the table, CEGHI. After simplification, there remain 7 sets of size 3:

CDJ, DEH, DEI, DEJ, DGJ, DHJ, DIJ

Character C is dependent on character B, so the first set is really BCDJ, and is not minimal. The remaining 6 sets all involve only vegetative characters, and are acceptable.

This method can be carried out by pencil and paper, but even an example as small as that in Fig. 1 (13 taxa, 18 characters) is very laborious, and so the use of a computer is advisable (Willcox and Lapage 1972). Even in a computer the amount of time and storage required increases rapidly with the size of the initial data matrix. This method would need modification if a minimal test set were wanted which guaranteed a given number of distinguishing characters for each taxon pair. First, each table entry would initially have to have at least as many characters as are wanted. Secondly, all characters which occur in entries with only the least desired number of distinctions would have to occur in the final minimal set.

Approximate methods for finding a minimal character set, i.e. one which is reasonably small but not necessarily the smallest, make use of a figure of merit for finding the best characters; for this the separation coefficient is very suitable since it directly measures the number of pairs of taxa which are separated. One method which has already been referred to is, in the program for identifying bacteria by maximum likelihood (Willcox *et al.* 1973) and is essentially that originally proposed by Gyllenberg (1963). When the identification score is inadequate, the program suggests further tests (characters) which ought to be carried out. It finds these by taking a selection of taxa from the list in order of the highest scores until the total combined score is sufficient (99.9 out of 100). Possible tests are those which have not already been used and which separate at least some pairs of taxa. To this purpose probabilities of test outcomes are approximated as positive or negative only. Tests are then selected in such a way that the next test taken is the one which separates the most additional pairs of taxa, while ignoring contributions to any pairs which are already separated by two or more tests. The addition of tests ceases when either all taxon pairs are separated by at least two tests or the available tests are exhausted. A similar program described by Rypka *et al.* (1967) finds minimal test sets by calculating the separation coefficient for several tests at a time. To begin with the first test chosen is that with the highest separation. This first test is then combined with each of the remaining ones to make a test pair and the separation calculated for all these pairs. The pair of tests with the best separation is then combined with a third, and so on, until a complete set is found which separates all taxa. This method is a compromise between searching every possible combination, which would take an enormous amount of computer time, and simply selecting tests in the order of their separation, which would ignore important correlations between tests. The statistical methods of character set reduction, discriminant analysis (p. 67) and principal co-ordinate analysis (Gower 1966) are loosely appropriate for finding minimum character sets but do not actually guarantee that every taxon pair will be distinct. The character weights calculated by these methods are analogous to separation coefficients and indicate which characters are likely to be needed in a minimal set.

An exact method to find diagnostic character sets is a simplification of the Kautz method described above. Instead of taking a table with all pairs of taxa, one need only consider those pairs involving the taxon of interest and apply the same procedure.

An approximate method for finding good diagnostic characters was described above in the context of an on-line identification program (p. 72). Another method was described by Barnett and Pankhurst (1974). To find a diagnostic description for a given species, begin with the complete description of the species with all characters and then try removing characters one at a time. If the taxon remains distinct from all other taxa, that character can be dropped entirely; otherwise it must be retained. Different results will be produced according to the order in which characters are dropped. Hence, characters were tested for rejection in reverse order of their separation coefficient, worst first and best last. Care is needed with variable conditional characters, as described above, which may not always effect a separation.

3.7 Continuous classification and identification

Since the matching methods of identification by computer use fairly complete descriptions of characters of specimens which are prepared for input to the computer, this provides an opportunity to refine the original data base with the use of accumulated information on fresh material. This emphasizes the fact that classification and identification are processes which form part of a repeating cycle. Roughly, the pattern of events is that, initially, a variety of organisms is collected and a classification is made from them, giving rise to keys or other means of identification. Subsequently, new material comes to light which cannot be identified satisfactorily, or new biological evidence comes to light, or both, and the classification has to be revised, giving rise to corrected identification procedures, and so on. This kind of cycle has been going on for centuries, but in some recent cases where computers have been involved the time scale has shrunk remarkably. This is particularly noticeable in the classification of bacteria which harm human beings, see e.g. Rypka (1975), who coined the phrase *continuous classification and identification* to describe the cycle.

One cannot however simply take the description of a newly identified specimen and incorporate it into the original data base without further ado. Assuming that the identification is correct, the specimen description needs to be checked in case any of the characters were wrongly observed. This is one reason why both the matching and maximum likelihood programs described above report on unusual character states in specimens which are positively identified. Even when unexpected character states have been checked, it may still be unwise to add all such new evidence to the data base, unless sufficient cases of the new varieties have been observed. Otherwise, the tendency in the long run may be to obscure useful distinctions between taxa by accumulating records of rare and aberrant varieties.

3.8 Special techniques for identification

The background to this section has been given in Chapter 1 (p. 9). Any technique which gives rise to a reproducible graph on a data recorder or a consistent black and white pattern which can be digitized with a television camera might be suitable. Possibilities include chromatography in one or two dimensions, stained electrophoretic gel patterns, spectrophotometry, mass spectroscopy and neutron-activation analysis. Direct analysis of photographs is not promising unless the objects concerned are flat or symmetrical and simple in nature, e.g. insect wings, colonies of bacteria. More details are given in a review by Morse (1975). These methods show promise but are not yet ready for general use.

3.9 Comparison of methods

This section is intended as a guide in selecting a suitable method for an identification problem. No single method is suitable for all purposes. The simplest methods are the cheapest and easiest to operate and are also the least flexible and most error-prone. Conversely, the complex methods are more costly and complicated to use and are the most flexible and reliable. The simpler methods are adequate for small numbers of highly distinct taxa but something more sophisticated may be required when the taxa are numerous and closely alike. If the taxa overlap extensively, it is better to use populations and statistical methods. Little or no information is available in the relative cost and effectiveness of the different methods applied to the same identification problem.

Fig. 18 compares the different methods on various criteria. The complexity of the methods increases from left to right, roughly speaking. A distinction is made between whether a computer is needed to set up the method on the one hand, or needed to operate it on the other. For those which can or must be operated by computer the usual mode

Fig. 18. Comparison of identification methods.

	Diagnostic key	Punched card key	Matching		
			Similarity	Bayesian/ max. like.	On-line
Construction by computer	Yes or no	Yes or no	—	—	—
Needs a computer to use	No	No	Yes (batch) or no	Yes (batch)	Yes (on-line)
Can be used in the field	Yes	Yes	Yes or no	No	No
No. of characters needed per specimen	Few	Few	Many	Many	Variable
Works with fragmentary specimens	No	Yes	Yes	Yes	Yes
Result if some characters of specimen in error	Often wrong	Often wrong	Usually right	Usually right	Usually right
Can detect new taxon	Usually not	Usually not	Yes	Yes	Yes
Numerical estimate with answer	No	No	Yes	Yes	Yes

of operation (batch or on-line) is given. Whether or not the method can be used in the field is shown next and this is just the opposite of the last criterion. The number of characters needed refers to whether the method is one of step-by-step elimination (monothetic) where relatively few characters are needed from each specimen, or whether it is one of comparison (polythetic) where it is better to describe the specimen completely. The on-line method can approach either extreme according to how it is used. Many biologists, such as palaeontologists or forensic scientists, are faced with fragmentary material where characters are missing at random, therefore need a multi-access identification method which allows them to use whichever characters are available. Curiously, the one method which is not suitable in these circumstances is the one which is currently the most popular—the diagnostic key. While the effect of errors has been discussed as appropriate under each particular method, the vulnerability of each method to errors of character description in specimens is much greater for monothetic methods than for polythetic ones. Only some of the methods give a numerical estimate of the correctness of the indicated identification, a consideration which is sometimes important. Correspondingly, those methods with a numerical estimate can show when the specimen is a member of some taxon other than those included. This error situation can pass undetected in step-by-step elimination methods.

3.10 Special kinds of specimen

Hybrids between species occur in nature and their identification presents difficulties. If a hybrid occurs frequently and has a characteristic appearance, it can be treated as a separate taxon and described like the rest. If many hybrids are possible in a taxonomic group, and if they are fertile and backcross further, the situation can be too difficult to cater for by adding extra taxa. A hybrid presented to a diagnostic key could then easily fail to key out, or key out incorrectly. A matching procedure may do better because one or both of the parents can be expected to appear in the resulting list of taxa. This is because many hybrids are intermediate between the parents, or resemble one parent more closely than the other because of dominant and recessive genes. If there is such a degree of interbreeding that all kinds of intermediates occur, it may be necessary to lump both parents with the hybrids in one aggregate taxon in order to distinguish the whole from other related taxa. If the problem is only to distinguish a possible hybrid from its parents where all three may occur in the same locality, a *hybrid index* can be calculated. This is a similarity coefficient based on that set of characters in which the parent taxa differ.

A similar difficulty occurs in medical diagnosis, but is not analogous to the hybrid in biology. This is *multiple pathology*, where the patient is suffering from two or more diseases simultaneously. The number of possible combinations is great, so it is not practical to describe them all. By and large all symptoms and signs (characters) of each disease occur, unless substantially the same parts of the body are involved, when they may interact or mask one another. The diagnosis has to account for all characters and this can be done by starting, fairly arbitrarily, with particular prominent characters and finding which diseases can account for these, until all characters are covered. One way to do this would be with a set of diagnostic keys, or equivalent, so that each key tackles a prominent character, e.g. fever, headache, and is cross-referenced to the other keys. Experience would still be needed in order to group the characters correctly. A large single key to all diseases would be much less helpful.

Finally, there is the problem of mixtures. This is encountered by pharmacognosists who may be presented with a powdered drug which is an intimate mixture of products derived from several different plants and asked to name the constituents. There is not much that can be done in this situation apart from skilled guesswork, unless some of the characters seen in the mixture are uniquely diagnostic for particular plant products.

4 History of Identification Methods

This chapter is a brief outline of the history of the subject. Voss (1952) reviews it in more detail.

It is hard to say when mankind first began to classify plants and animals. The earliest 'folk' taxonomy was probably a very simple one with two taxa only: 'useful' and 'not useful'. Under 'useful' came anything which could provide food, medicine or shelter, and under 'not useful' came poisons, pests, weeds and everything else. Presumably recognition was by familiarity alone. The earliest written taxonomies were clearly hierarchical but there is no evidence that these were consciously used in the manner of keys. A typical passage from Aristotle, Historia Animalium, Book I, chapter 1, reads:

> 'Of land animals some are furnished with wings, such as birds and bees, and these are so furnished in different ways one from another; others are furnished with feet. Of the animals that are furnished with feet, some walk, some creep, and some wriggle.'

Theophrastus, pupil of Aristotle, prepared a detailed classification of plants and lists many binary characters. There followed a very long period of what, to modern eyes, seems to have been more or less complete stagnation in the study of natural history in the western world. Scholars were largely content to copy from older sources and to use their imagination rather than make observations of their own. Herbals and bestiaries of the middle ages were often illustrated in a fanciful way but must have been used to some extent for identification of specimens by comparison with the drawings.

In the 17th century a number of works appear where classifications are set out in diagrammatic form. One of the most striking is due to Morison (1672), discussed and illustrated by Walters (1975), as part of a classification of the *Umbelliferae*. Morison has drawn what is clearly a hierarchical tree (cf. Fig. 3) but does not state that it can serve as a key. Similarly Ray (1686), in the *Historia Piscium*, published a 'table' of the cartilaginous fish (Fig. 19) which looks just like a key, but again does

Fig. 19. Table of the cartilaginous fishes (due to John Ray).

PISCIUM
CARTILAGINEORUM
Tabula.

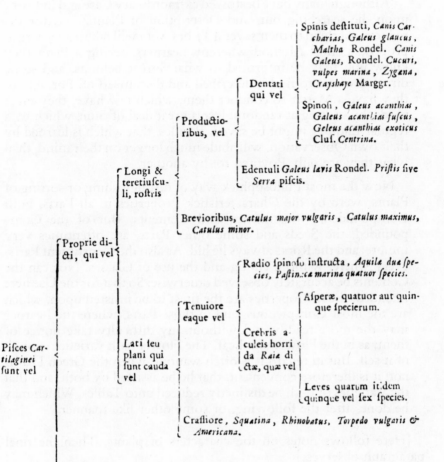

Pisces *Cartilaginei* sunt vel

Proprie dicti, qui vel

— Longi & teretiusculi, rostris

— Productioribus, vel

— Dentati qui vel

Spinis destituti, *Canis Carcharias, Galeus glaucus, Maltha* Rondel. *Canis Galeus*, Rondel. *Cucuri, vulpes marina, Zygæna, Crayshaye* Marggr.

Spinosi , *Galeus acanthias , Galeus acanthias fuscus , Geleus acanthias exoticus* Clus. *Centrina.*

Edentuli *Galeus lævis* Rondel. *Pristis* sive *Serra* piscis.

Brevioribus, *Catulus major vulgaris , Catulus maximus, Catulus minor.*

Lati seu plani qui sunt cauda vel

— Tenuiore eaque vel

Radio spinoso instructa , *Aquila duæ species, Pastinæca marina quatuor species.*

Asperæ, quatuor aut quinque specierum.

Crebris aculeis horrida *Raiæ* dictæ, quæ vel

Leves quarum itidem quinque vel sex species.

Crassiore , *Squatina, Rhinobatus, Torpedo vulgaris & Americana.*

Minus proprie dicti , qui pro ossibus cartilagines habent , verum nec 5. illas scissuras ad branchias obtinent, nec mentulas pinnis ad anum subnexas, nec pariter ovipari sunt & vivipari.

LIB.

not state that it is such. By contrast, Grew (1682) states quite clearly how to approach the writing of a key but does not actually carry out his own instructions. He comments on the need for a key and gives advice on choice of characters, as follows:

'Although many have bestowed extraordinary Care and Industry upon the searching out, and Description of Plants; and for the reducing of them to their several Tribes: yet I will take leave, here to propose a short Method whereby Learners, seeing a Plant they know not, may be informed to what Sort it belongs, and so be directed where to find it described and discoursed of. For, except they have a Matter to conduct them, which few have; they must needs, by seeking at random, lose a great deal of time, which by a regular Enquiry might be saved. Besides, that which is learned by their own Observation, will abide much longer on their mind, than what they are only Poynted to, by another.

Now the most Philosophick way of distinguishing or sorting of Plants, were by the Characteristick Properties in all Parts, both Compounded, Constituents, and Contents. But of the Compounded, the Seeds and some other Parts, are oftentimes very minute: and the Roots always lie hid. As also the Constituent Parts, every where, without cuting and the use of Glasses. Nor can the Contents be accurately observed otherwise. So that for the Use here intended, those Properties are the fitest to be insisted upon, which are the most Conspicuous, and in those Parts, where the Learner may the most readily and without any difficulty take notice of them; as in the Flower and Leaf. The Flower hath varieties enough of it self. But in regard it is often wanting when the Green Leaf is not; it is therefore convenient, that he be assisted by both, and that the Varieties of both be distinctly reduced unto Tables. Which may be done, after the following, or some other like manner.'

[Here follows notes on the characters of plants. Then the final paragraph observes:]

'How far these, and some other like Distinctions, being reduced to Tables, would serve for the finding out of any Sort of Plant, may be conceived, if we consider, how great a Variety, a few Bells, in the ringing of Changes, will produce. And the search will be easy, and successful, if in every foregoing Table, reference be made to those that follow; and the Tables conteining the last Divisions, the Names of Plants therein poynted out, be expressed.'

Linnaeus, the 'father of modern taxonomy', described a key in 1736 and was the first to call it such (*clavis*, in Latin), but curiously his key was to botanists, not to plants! Nearly a full century passed after Grew's publication before the appearance of what are quite unmistakably keys, in the modern sense, in Lamarck's *Flore Françoise* (Flora of France) in 1778. Part of one of his keys is shown in Fig. 20. Lamarck describes how the keys are constructed and used, and understood their basic principles (cf. p. 18). In his own words (Vol. 1, introduction, p. lxxiii):

'Je dois observer ici que la manière de procéder dans une analyse ne peut être arbitraire, et qu'encore qu'il paroisse indifférent au premier coup-d'oeil d'employer telle division plutôt que telle autre, la marche, qui fera trouver le nom de la plante, doit cependent être combinée d'après certaines règles que je réduis à deux. La première est que l'on parvienne au but par la voie la plus sûre. La seconde est que cette voie soit en même-temps la plus courte possible.'

[I must remark here that the method of proceeding in an analysis (identification) cannot be arbitrary, and that although it may seem unimportant at first glance whether one division (character) rather than another is used, the path which will lead to the name of the plant has nevertheless to be constructed according to certain rules, which I reduce to two. The first is that one must arrive at the result by the most reliable route. The second is that this route should, at the same time, be the shortest possible one.]

In the two centuries since Lamarck keys of various kinds have become the most popular means of specimen identification among biologists. A few additional ideas have been suggested in the meantime, such as keys with backward references (Bonnier 1917), and the self-indexing variety due to Evans (1949) already referred to (p. 23), but the basic idea remains the same.

The 20th century has seen most of the innovations in identification methods. The idea of the tabular key, called by its author a 'synoptic key', appears with Ogden (1943) and Gyllenberg seems to have been the first to try out a matching type of program on a computer for a biological application (see Chapter 3). Cards for edge-punching were patented in the 1920s and the first applications to making keys appeared in the 1930s, e.g. Clarke (1937). The word 'polyclave' was first coined by Duke (1969) who used it for a card overlay scheme which is similar to the body-punched card key, for which the term is now often used. The first on-line identification program appears to have been that of Boughey *et al.* in 1968 (see Chapter 3).

The application of computers to identification began in the early 1960s, some 10 to 15 years after the beginnings of computers themselves (usually dated as 1947) but quite soon after the appearance of

ANALYSE.

| Fleurs dont les étamines & piſtils peuvent aiſément ſe diſtinguer. 1. | Fleurs dont les étamines & piſtils ſont nuls, ou ne peuvent ſe diſtinguer. 16. |

1.

Fleurs dont les étamines & piſtils peuvent aiſé-ment ſe diſtinguer...
{ Fleurettes nombreuſes, réunies dans un calice com-mun............. 2.
Fleurs libres & non réunies dans un calice commun.. 9

2.

Fleurettes nombreuſes, réunies dans un calice commun..........
{ Fleurettes de même ſorte; elles ſont toutes en cornet, ou toutes en languette..... 3
Fleurettes de deux ſortes, les unes en cornet, & les autres en languette..... 6

3.

Fleurettes de même ſorte..........
{ Fleurettes toutes en cornet.............. 4
Fleurettes toutes en lan-guette............. 5

4. Fleurettes toutes en cornet.

Carduus marianus.

5. Fleurettes toutes en languette.

Hieracium murorum.

e iij

general-purpose high level computer languages for general use (late 1950s), which was probably a necessary prerequisite. It may well be true to say that most of the basic techniques for identification with the aid of a computer have now been worked out, and that we are now in a period of expansion of applications. The use of computers makes possible a very much wider range of different methods for different purposes and a much greater flexibility than the traditional techniques could offer.

Fig. 20. Diagnostic key to higher plants (Lamarck).

5 Applications

This chapter will give some details of the kind of identification problems encountered in different branches of biology with some comment on medical diagnosis also, which is a rather similar problem. It is not intended as a guide to the principal works for the identification of the world's flora and fauna, although some such publications will be mentioned. Guides for this purpose already exist, e.g. Kerrich *et al.* (1978). Nor will it review the many techniques available for obtaining information about specimens, which range from direct observation to the use of complex laboratory procedures and large-scale electronic equipment; details of these are to be found in the appropriate standard taxonomic textbooks. Computer applications will be dealt with in their systematic context.

The following remarks apply to most taxonomic groups. The availability of means of identification varies according to the state of taxonomic knowledge of the group concerned. Where the taxonomy is in an early stage, there is a tendency for little to be available and that to be unreliable, e.g., for protozoa. Conversely, where taxonomy is more advanced, e.g. for higher plants in temperate regions, a wide variety of good keys is on hand. There are regional differences as well, for example, the temperate zones are far better studied than the tropical lowlands. Also, the degree to which different groups have been studied is related more to the ease with which they can be studied, their popular appeal and their economic or practical importance to man, rather than to their diversity or biological or ecological significance. The overwhelming majority of identification aids published are diagnostic keys of one form or another. The reasons for this must be their historical precedence, the practicality of keys in the field, and the ease with which they can be published and distributed. Recent technical innovations provide other ways of making identifications which are also practical in the field and are easily distributed, but which avoid the drawbacks of diagnostic keys, so this situation may be expected to change. Finally, it must be said that there occur instances in all parts of biology where there are numerous taxa to distinguish and although these may be all quite distinct, it is more effective to provide a good set of drawings or photographs than to describe the differences in words.

One difficulty which besets taxonomists of all kinds is the distortion or loss of information caused by the preservation of specimens. The problems include fading of colours, distortion of shape, lack of adequate field notes, and the loss or decay of soft or fragile parts. This is sometimes allowed for by providing keys which are especially intended for preserved specimens.

5.1 Botany

Botanists are generally well provided with keys, especially for vascular plants in temperate zones. There is less adequate provision for cryptogamic plants, the smaller fungi and tropical flora. There is a strong emphasis on mature (flowering or fruiting) material, and one would have much more difficulty in trying to name seeds, seedlings, or immature plants. Since both sexes usually occur on the same specimen, separate keys to male and female plants are not often needed. Computer-constructed diagnostic keys have begun to appear, e.g. Watson and Milne (1972). The virtues of punched card keys have long been realized, for example the key to angiosperm families of Hansen and Rahn (1969) (see p. 30). Another striking example is a very extensive edge-punched key to wood samples (Phillips 1948), which is particularly intended for fragments. An example of a computer produced card key is that by Weber and Nelson (1972). Tabular keys have already been quoted, such as that of Sinker (1975) (p. 33) and Ogden (1943) (p. 85). Matching programs using similarity have also been discussed (p. 55), and there is a botanical example in Pankhurst (1975). Palynologists have attempted to identify pollen grains in a similar way (Walker *et al.* 1968). Baum and Lefkovitch (1972) applied the Bayesian method (p. 62). The original on-line identification program (Boughey *et al.* 1968) (p. 68) was applied to the flora of Orange County, California.

5.2 Zoology

It is said that three-quarters of the biological species in the world are insects. However, this is not to say that entomologists and entomological identification keys occur with the same relative abundance. Taking zoology generally, the diagnostic key is again the favourite method but zoologists are probably not so well provided for as botanists. It seems that the difficulties of making descriptions are greater and keys are frequently illustrated, e.g. Fig. 6. As a further example, the monumental, many-volume *Tierwelt Mitteleuropas* (Animal World of Central Europe) of Brohmer *et al.* (1960) contains numerous keys and illustrations. Quite often the standard reference works to animal groups simply contain illustrations and descriptions with few or no diagnostic keys, e.g. the FAO series on commercial fishery (1973) or one of the standard works on protozoa (Kudo 1966). There is something of a tradition in some groups, e.g. butterflies and moths, birds, to have only volumes of descriptions and pictures and not to use keys at all. It may also be fair to say that zoologists (other than entomologists) have more difficulty in preserving specimens for study than do botanists. The need for different keys to males versus females, and larvae versus adults occurs frequently.

Entomological keys have been constructed by computers, e.g. Dallwitz (1974), and tabular keys have been put forward (Newell 1970). Examples of the use of discriminant analysis in zoology are given by Sneath and Sokal (1973). It seems that there are not as yet as many examples of zoological identification by computer as there are in other disciplines.

5.3 Palaeontology

This subject presents some contrast to botany and zoology. The palaeontologist is frequently faced with fragmentary specimens, especially of larger vertebrates. There is also the effect of accidents of preservation, where specimens are crushed or distorted, and where some of the characters which appear to be present really derive from the rock rather than from the original organism. The general custom is to compare specimens with collections or illustrations rather than to use keys. There are a number of impressive general publications which are well illustrated, but which contain few or no keys, such as the many-volume *Treatise on Invertebrate Paleontology* (Geological Society of America), and the catalogues for the Foraminifera (Ellis and Messina 1940), and for the Ostracoda (Sylvester-Bradley and Siveter 1973), where the illustrations are photographs in stereo pairs. There are exceptions to this general rule, such as the keys to fossil corals (Cotton 1973) and in palaeobotany (Harris *et al.* 1961).

If the reasons for the unpopularity of diagnostic keys in this area is the problem of fragmentary specimens, there is much scope here for the application of the other methods. The only computer application known at this moment is a data base on fossil pollen grains (Germeraad and Muller 1972). In this case, nonetheless, specimens are generally complete, if distorted. This data base is used for identification by searching via information retrieval techniques, which is a form of matching by similarity.

5.4 Microbiology

Microbiologists can preserve their specimens as can other taxonomists but they cannot 'see' their specimens in the same way as the characters are not self-evident as they are, say, on a herbarium sheet or in a spirit jar. The characters are mostly chemical tests, taking a long time to complete, and so the results usually get coded onto record sheets. There are few or no morphological characters. Given the amount of idenfication work in bacteria which is carried out at public health laboratories, large amounts of descriptive specimen data accumulate quickly. These are probably the reasons why bacteriologists began to use computer methods before other taxonomists, both in classification and identification.

Because the tests take a long time, they are carried out in batches. If identification is attempted with a diagnostic key, and a character (test) which is requested has not been done, one is in much the same situation as any other biologist with a fragmentary specimen. If the specimen comes from someone who is ill in hospital, there may not be time for more tests to be completed to establish the identification before treatment can begin. On the other hand, the tests cost money and it is not reasonable to carry out all tests on all specimens. In addition, bacteria often occur in atypical 'strains' as a result of relatively rapid evolution, and for all these reasons multi-access or polythetic identification methods seem to be preferred.

Conventional diagnostic keys for bacteria seem to be few but there have been many computer applications. Punched card keys have been prepared and one, the Pathotec Rapid Identifier produced by General Diagnostics Inc. in 1975, is commercially available. A mechanical polyclave for bacteria (Olds 1970) was described on p. 32. In the class of matching methods, the work of Gyllenberg (1963) and Gyllenberg and Niemelä (1975) has been discussed (p. 55), and the program of Willcox et al. (1973) has been described in detail (p. 64). These methods often use the maximum likelihood approach because of the availability of data on the probability of the occurrence of character states. A considerable number of related papers has appeared since 1970.

There has also been some work on yeasts which are very useful in brewing and food manufacture. Barnett and Pankhurst (1974) have a lengthy computer-constructed diagnostic key, for which an equivalent punched card key exists and a set of computer-derived diagnostic descriptions. A matching procedure for computer identification of yeasts is described by Campbell (1973).

5.5 Pharmacognosy

Pharmacognosy is the study of natural products (mostly from plants) giving rise to pharmaceutical products. Pharmacognosists are often asked to identify the raw materials of drugs, usually powdered plant material, or root segments or fruit. When the material is a mixture the identification problem is severe (p. 80). There exist illustrated textbooks showing the microscopic details of such materials, e.g. starch grain type, oxalate crystal shape. One difficulty is the unrestricted variety of plant products which may be encountered, so that it is hard to make a key complete. There is however, one large diagnostic key (Claus 1956) and a punched card key (Nelson 1972). Joliffe and Joliffe (1976) describe a computer program for identifying drug materials which uses matching by similarity.

5.6 Medical diagnosis

Medical diagnosis is not, of course, a biological problem except in the widest sense, but it is very interesting to compare it with biological identification.

A disease is not strictly analogous to a biological taxon because there is not a similar body of theory to explain why it is separate from other diseases and how it maintains this separation. On the other hand diseases are in fact largely distinct from each other; if they were not, there could be no diagnosis. Whether a disease is something which exists in its own right, or whether it is simply a convenient label for the correspondence between a set of symptoms and a course of treatment, is not a question which has to be answered before diagnosis and treatment can be carried out. However, many diseases have known causes, such as infection by bacteria or some bodily malfunction, which can often explain the symptoms. In fact the 'characters' of a disease are of three kinds: (i) symptoms, which the patient is aware of and may complain of; (ii) signs, which the doctor observes but which the patient may not know about; and (iii) tests, often chemical and quantitative, usually carried out in a laboratory on a sample. These distinctions do not affect the logic of diagnosis but tests tend to be more expensive or inconvenient for the patient.

Human diseases are very few in number (thousands) compared with biological species (millions). A naïve and ignorant biologist might expect to find that these diseases were all thoroughly classified and described in textbooks of medicine, along with suitable keys for recognizing them. Certainly there are medical textbooks which describe diseases but their classification, compared with biological classifications, is rather unfinished. There are good reasons for this; one is that taxonomic studies are not the most urgent priority for the medical profession; another that diseases are quite high-level taxa, corresponding, say, to the level of the family rather than the species, and include a great deal of variation. While the development of a disease from its beginning to its climax corresponds to the changes in an organism from birth to maturity, there is nothing biological which really corresponds to the variation in disease characters brought about by different stages in different possible courses of treatment.

Doctors mostly seem to be taught to diagnose by looking at examples rather than by following an explicit logical procedure. In a survey, Freemon (1972) found that more than half the medical staff questioned thought that they diagnosed by some kind of step–by–step elimination of diseases by characters, whereas others thought that they did some sort of matching or assessed probabilities of different diseases. There certainly exist diagnostic characters which are specific, or nearly so, to particular diseases, e.g. the spasm of the jaw muscles seen in tetanus, but only some diseases can be recognized by such means. There is, therefore, more or less nothing in the conventional practice of medical diagnosis to correspond to the use of diagnostic keys by biologists.

The first applications of computers to medical diagnosis appeared at about the same time as biologists began to apply computers to identification problems. An important instance is the program of Brodman *et al.* (1959) which was a form of matching by similarity with character weighting. Most of the programs since then have employed Bayes' theorem or maximum likelihood, as described in the review by Croft (1972), where a plea is made for a concerted effort to compile a definitive set of standardized disease descriptions, which scarcely yet exists, and for study of the comparative effectiveness of different methods. A biologist might be surprised to learn that the diagnostic key and the punched card key, or other forms of polyclave or tabular keys have scarcely been tried in medical diagnosis. Notable exceptions are the logoscope (Nash 1960) (see p. 35 and Fig. 12) and the diagnostic flowcharts of Essex (1975). The diagnostic flowcharts are logically the same as diagnostic keys, except that they are drawn in the form of decision trees, as the key in Fig. 3, with the questions (characters) in boxes joined to other boxes by arrows. These charts were prepared for use in rural communities in Tanzania and were given field trials to ensure accuracy. The use of many different flowcharts, each for a prominent symptom or symptom group with cross-references, is a solution to the problem of multiple pathologies referred to earlier (p. 80). Hence one may say that all the various methods have yet to be fully tried out in medical diagnosis, as is also true in biology but the pattern of use and development has so far been very different.

5.7 Summary

The previous discussion has not covered every possible area of application for identification techniques in biology but an effort has been made to cover all the kinds of problem which can occur. In the author's view, both the problems and their solutions are more uniform than is generally supposed to be the case.

References

BARNETT, J. A. & PANKHURST, R. J. (1974). *A New Key to the Yeasts*. North-Holland, Amsterdam, 273 pp.

BAUM, B. R. & LEFKOVITCH, L. P. (1972). A model for cultivar classification and identification with reference to oats (*Avena*). II. A probabilistic definition of cultivar groupings and their Bayesian identification. *Canadian Journal of Botany*, **50**, 131–138.

BONNIER, G. (1917). *Les Noms des Fleurs par la Méthode Simple*. Paris, Librairie Générale, 332 pp.

BOUGHEY, A. S., BRIDGES, K. W. & IKEDA, A. G. (1968). *An Automated Biological Identification Key*. Museum of Systematic Biology, University of California, Irvine. Research Series No. 2, 36 pp.

BRODMAN, K., VAN WOERKOM, A. J., ERDMANN, A. J. & GOLDSTEIN, L. S. (1959). Interpretations of symptoms with a data-processing machine. *Archives of Internal Medicine*, **103**, 116–122.

BROHMER, P., EHRMANN, P. & ULMER, G. (1960). *Tierwelt Mitteleuropas*. In many volumes. Quelle & Meyer, Leipzig.

BRUES, C. T., MELANDER, A. L. & CARPENTER, F. M. (1954). *Classification of Insects*. Harvard University Press, 917 pp.

CAMPBELL, I. (1973). Computer identification of the genus *Saccharomyces*. *Journal of General Microbiology*, **77**, 127–135.

CHINERY, M. (1973). *A Field Guide to the Insects of Britain and Northern Europe*. Collins, London, 352 pp.

CLAPHAM, A. R., TUTIN, T. G. & WARBURG, E.F. (1962). *Flora of the British Isles*, 2nd edition. Cambridge University Press, Cambridge & London, 1269 pp.

CLARKE, S. H. (1937). *The Construction of Keys to the Identification of Timber*. Forest products Research Lab, Progress Report No. 4, 19 pp.

CLAUS, E. P. (1956). *Gathercoal and Wirth's Pharmacognosy*, 3rd edition. Henry Kimpton, London, Appendix p. 687.

COTTON, G. (1973). *The Rugose Coral Genera*. Elsevier. Amsterdam, 358 pp.

COWAN, S. T. & STEEL, K. J. (1960). A device for the identification of micro-organisms. *Lancet*, **1**, 1172–1173.

COWAN, S. T. & STEEL, K. J. (1965). *Manual for the Identification of Medical Bacteria*. Cambridge University Press, Cambridge & London, 217 pp.

CROFT, D. J. (1972). Is computerised diagnosis possible? *Computers and Biomedical Research*, **5**, 351–367.

DALLWITZ, M. J. (1974). A flexible computer program for generating diagnostic keys. *Systematic Zoology*, **23**(1), 50–57.

DAVIS, P. H. & CULLEN, J. (1965). *The Identification of Flowering Plant Families.* Oliver and Boyd, Edinburgh.

DUKE, J. A. (1969). On tropical tree seedlings I. Seeds, seedlings, systems and systematics. *Annals of the Missouri Botanical Garden,* **56**(2), 125–161.

ELLIS, B. F. & MESSINA, A. (1940 *et seq.*). *Catalogue of Foraminifera.* American Natural History Museum, New York.

ESSEX, B. J. (1975). *Diagnostic pathways in clinical medicine.* Churchill Livingstone, Edinburgh, 176 pp.

EVANS, W. H. (1949). *A Catalogue of the Hesperidae.* British Museum (Natural History), London, 502 pp.

F. A. O., (1973). *Species Identification Sheets for Fishery Purposes. Many volumes.* FAO, Rome.

EINSTEIN, A. (1958). *Foundations of Information Theory.* McGraw Hill, New York, 135 pp.

FREEMON, F. R. (1972). Medical diagnosis, comparison of human and computer logic. Biomed. comp. **3**, 217–221.

GERMERAAD, J. H. & MULLER, J. (1972). *Computer-based Numerical Coding System for the Description of Pollen Grains and Spores.* National Museum of Geology & Mineralogy, Leiden. 2 vols.

GOWER, J. C. (1966). Some distance properties of latent root and vector methods used in multivariate analysis. *Biometrika,* **53**, 325–338.

GOWER, J. C. & PAYNE, R. W. (1975). A comparison of different criteria for selecting binary tests in diagnostic keys. *Biometrika,* **62**(3), 665–672.

GREW, N. (1682). Appendix to Part 2 of Book 4 of *The Anatomy of Plants with an Idea of a Philosophical History of Plants. . . .* W. Rawlins (printer), pp. 174–176.

GYLLENBERG, H. G. (1963). A general method for deriving determination schemes for random collections of microbial isolates. *Annals of the Finnish Academy of Sciences, ser. A, IV. Biology,* **69**, 1–23.

GYLLENBERG, H. G. & NIEMELÄ, T. K. (1975). New approaches to microbial identification. In *Biological Identification with Computers* (ed. R. J. PANKHURST), pp. 121–136. Academic Press, London & New York.

HALL, A. V. (1973). The use of a computer-based system of aids for classification. *Contr. Bolus Herb,* **6**. University of Cape Town, 110 pp.

HANSEN, B. & RAHN, K. (1969). Determination of angiosperm families by means of a punched-card system. *Dansk Botanisk Arkiv.,* **26**, 46 pp. & 172 punched cards.

HARRIS, T. M. (1961). *Yorkshire Jurassic Flora,* 4 volumes. British Museum (Natural History), London.

HENNIG, W. (1966). *Phylogenetic Systematics.* University of Illinois Press, Urbana, 263 pp.

JOLIFFE, G. H. & JOLIFFE, G. O. (1976). Computer-aided identification of powdered vegetable drugs. *Analyst,* **101**, 622–633.

KAUTZ, W. H. (1968). Fault testing and diagnosis in combinatorial digital circuits. *IEEE Trans. Comput,* **C17**, 352–366.

KERRICH, G. J., HAWKSWORTH, D. L. & SIMS, R. W. (1978). *Key works to the Fauna and Flora of the British Isles and Northwestern Europe.* For Systematics Association by Academic Press, London & New York.

KOŽUHAROV, S. (1976). *Jurinea*. In *Flora Europaea* (ed. T. G. TUTIN *et al.*), vol. 4, pp. 218–220. Cambridge University Press, Cambridge & London.

KUDO, R. R. (1966). *Protozoology*, 5th edition. C. C. Thomas, Springfield, Illinois, 1174 pp.

LAMARCK, J. B. P. (1778). *Flore Françoise*, 1st edition. 3 vols. Paris, Imprimerie Royale.

LAWRENCE, G. H. M. (1951). *Taxonomy of Vascular Plants*. Macmillan, New York, 832 pp.

LINNAEUS, C. (1736). *Clavis Classium in Systemate Phytologorum*, in Bibliotheca Botanica, Amsterdam.

MAYR, E. (1969). *Principles of Systematic Zoology*. McGraw-Hill, New York, 428 pp.

MENDEL, J. M. & FU, K. S. (eds) (1970). *Adaptive Learning and Pattern Recognition*. Academic Press, New York, 444 pp.

MINSKY, M. & PAPERT, S. (1969). *Perceptrons*, chapters 0 & 11. MIT Press, Cambridge, Mass.

MORISON, R. (1672). *Plantarum Umbelliferarum Distributio Nova* . . . Oxford, Sheldonian Theatre, 91 pp.

MORSE, L. E. (1974) Computer programs for specimen identification, key construction and description printing. *Publs. Mus. Mich. St. Univ., Biol.*, **5**, 128 pp.

MORSE, L. E. (1975). Recent advances in the theory and practice of biological specimen identification. In *Biological Identification with Computers* (ed. R. J. PANKHURST), pp. 11–52. Academic Press, London & New York.

MORSE, L. E., PANKHURST, R. J. & RYPKA, E. W. (1975). A glossary of computer-assisted biological specimen identification. In *Biological Identification with Computers* (ed. R. J. PANKHURST), pp. 315–330. Academic Press, London & New York.

NASH, F. A. (1960). Diagnostic reasoning and the logoscope. *Lancet*, **1**, 1442–1446.

NELSON, P. F. (1972). *Analytical Microscopy of Vegetable Materials*. Available from the author at 114, Corsebar Drive, Paisley, Scotland.

NEWELL, I. M. (1970). Construction and use of tabular keys. *Pacific Insects*, **12**, 25–37.

OGDEN, E. C. (1943). The broad-leaved species of *Potamogeton* of North America North of Mexico. *Rhodora*, **45**, 57–105, and Contr. Gray Herb. No. 147.

OLDS, R. J. (1970). Identification of bacteria with the aid of an improved information sorter. In *Automation, Mechanization and Data Handling in Microbiology* (eds A. BAILLIE & R. J. GILBERT), pp. 85–89. Academic Press, London & New York.

OSBORNE, D. V. (1963). Some aspects of the theory of dichotomous keys. *New Phytologist*, **62**, 144–160.

PANKHURST, R. J. (1971). Botanical keys generated by computer. *Watsonia*, **8**, 357–368.

PANKHURST, R. J. (1975). Identification by matching. In *Biological Identification with Computers* (ed. R. J. PANKHURST), pp. 79–91. Academic Press, London & New York.

PANKHURST R. J. (1976). *On-line Identification Program, Version 2.* Internal document. British Museum (Natural History), London.

PANKHURST, R. J. & AITCHISON, R. R. (1975). A computer program to construct polyclaves. In *Biological Identification with Computers* (ed. R. J. PANKHURST), pp. 73–78. Academic Press, London & New York.

PAYNE, R. W. (1975). Genkey: a program for constructing diagnostic keys. In *Biological Identification with Computers* (ed. R. J. PANKHURST), pp. 65–72. Academic Press, London & New York.

PHILLIPS, E. W. J. (1948). *Identification of Softwoods by their Microscopic Structure.* Forest Products Research Bulletin No. 22, HMSO, London.

RAY, J. (1686). *Historia Piscium* . . . Oxford, Sheldonian Theatre, 343 pp.

ROSS, G. J. S. (1975). Rapid techniques for automatic identification. In *Biological Identification with Computers* (ed. R. J. PANKHURST), pp. 93–102. Academic Press, London & New York.

RYPKA, E. W. (1971). Truth table classification and identification. *Space Life Sciences*, **3**, 135–156.

RYPKA, E. W. (1975). Pattern recognition and microbial identification. In *Biological Identification with Computers* (ed. R. J. PANKHURST), pp. 153–180. Academic Press, London & New York.

RYPKA, E. W., CLAPPER, W. E., BOWEN, I. G. & BABB, R. (1967). A model for the identification of bacteria. *Journal of General Microbiology*, **46**, 407–424.

SHULTZ, L. M. (1975). Program CDKEY. In *Biological Identification with Computers* (ed. R. J. PANKHURST), p. 306. Academic Press, London & New York.

SINKER, C. A. (1975). A lateral key to common grasses. *Bulletin of the Shropshire Conservation Trust*, **34**, 11–18.

SNEATH, P. H. A. & SOKAL, R. R. (1974). *Numerical taxonomy.* Freeman & Co., San Francisco, 573 pp.

SYLVESTER-BRADLEY, P. C. AND SIVETER, D. J. (eds) (1973). *A Stereo-Atlas of Ostracod Shells.* Dept. of Geology. Univ. of Leicester, vol. 1 *et seq.*

VOSS, E. G. (1952). The history of keys and phylogenetic trees in systematic biology. *Journal of the Scientific Laboratories, Denison University*, **43**(1), 1–25.

WALKER, D., MILNE, P., GUBBY, J. & WILLIAMS, J. (1968). The computer assisted storage and retrieval of pollen morphological data. *Pollen et Spores*, **10**, 251–262.

WALTERS, S. M. (1975). Traditional methods of biological identification. In *Biological Identification with Computers* (ed. R. J. PANKHURST), pp. 3–8. Academic Press, London & New York.

WATSON, L. & MILNE, P. (1972). A flexible system for automatic generation of special purpose dichotomous keys, and its application to Australian grass genera. *Australian Journal of Botany*, **20**, 331–352.

WEBER, W. A. & NELSON, P. P. (1972). *Random-access Key to Genera of Colorado Mosses.* University of Colorado Museum, Boulder.

WHALLEY, P. E. S. (1976). *Tropical Leaf Moths.* British Museum (Natural History), London, 194 pp.

WILLCOX, W. R. & LAPAGE, S. P. (1972). Automatic construction of diagnostic tables. *Computer Journal*, **15**(3), 263–267.

WILLCOX, W. R., LAPAGE, S. P., BASCOMBE, S. & CURTIS, M. A. (1973). Identification of bacteria by computer: theory and programming. *Journal of General Microbiology*, **77**, 317–330.

Index